D0267981

8

The Biotechnology Revolution

The Biotechnology Revolution
An International Perspective

Alan M. Russell
Department of International Relations and Politics
North Staffordshire Polytechnic

WHEATSHEAF BOOKS · SUSSEX

ST. MARTIN'S PRESS · NEW YORK

NORTH STAFFORDSHIRE
POLYTECHNIC
LIBRARY

First published in Great Britain in 1988 by
WHEATSHEAF BOOKS LTD
A MEMBER OF THE HARVESTER PRESS PUBLISHING GROUP
Publisher: John Spiers
16 Ship Street, Brighton, Sussex

and in the USA by
ST. MARTIN'S PRESS, INC.
175 Fifth Avenue, New York, NY 10010

British Library Cataloguing in Publication Data
Russell, Alan M.
 The biotechnology revolution : an
 international perspective.
 1. Biotechnology   2. Technology and
 international affairs
 I. Title
 660'.6   TP 248.2

ISBN 0-7450-0013-4

Library of Congress Cataloging-in-Publication Data

Russell, Alan M.
 The biotechnology revolution.
 Revision of thesis (ph.d) — University of Kent Canterbury.
 'Wheatsheaf books'
 Bibliography: p.
 Includes index.
 1. Biotechnology industries — International cooperation.
 2. Biotechnology — International cooperation. 3. Biotechnology
 industries — Government policy. 4. Biotechnology —
 Government policy.
 I. Title
 HD9999. B44 2R87 1988 338.4'76606 87-35590
 ISBN 0-312-01876-2

Typeset in Times 11/12pt by
Witwell Ltd, Liverpool
Printed in Great Britain by
Billings & Sons Ltd, Worcester

For my mother and in memory of my father

Contents

Figures

Preface

We live in a world of growing complexity. The demands of economic, social and political management are ever-increasing, and are evident at all levels of social organisation from local communities, through national and transnational levels, to the international sphere. Part of the complexity arises out of the increasing interlinkage of all these levels, such that political decision-making occurs in an environment of influences and constraints founded upon global awareness. Of course, there are differences between regions of the world in terms of development, political awareness, culture, resources and the like. Indeed, there is a case to describe the prevalent international structure as one of dominance and dependence, between centres of economic development and under-developed countries of the global periphery. Yet there is, at a minimum, an international ethos that stresses the desirability of continued economic growth and technological sophistication. This is certainly the case in the developed world, but is also a view by and large transferred to the elites of the developing countries. Political ideologies may differ in their strategies for achieving such growth, and may differ in policies of distribution, yet can agree on the need for maintaining a position at the forefront of science and technology. It is the latest international revolution in technology that is to be addressed in this book.

Much attention and excitement has surrounded bio-technology in the 1980s. Journalists, academics, science commentators and industrialists have all shown great interest. However, despite the international significance of the technology, few have applied an international perspective. There is scope for this in two ways. The future implications of biotechnology are clearly going to be international, affecting patterns of trade, technology transfer, agricultural inno- vations, and so on. This book will introduce some obser- vations in this context. However, a more specific objective of this study is to examine the history of what has already happened from an international orientation. In particular, by focusing on the contentious discussions of safety, the emergent policy frameworks will be shown to be highly interdependent transnationally. This extends even to the sharing of common errors and to the taking of sides in political posturing. As with the nuclear industry the past must not be forgotten, as we live with the present consequences. The tendency continually to look to the future in an adolescent technology should, there- fore, be moderated.

It is hoped that a wide audience will find some appeal in this work. Although my training is in the field of international relations my orientation is quite interdisciplinary. Anyone with an interest in public policy, political science, sociology and, not least, technology itself, should find something of relevance to their specialisms. However, I must also be honest. Implicit in this study is a criticism of specialisation itself, as traditionally categorised. This extends from scientific 'experts' addressing safety, to academics enclosed too narrowly within disciplines. The impact of technology affects all levels and aspects of society and we compartmentalise its study at the expense of highlighting the common links. That said, this work like all academic writings is designed to elicit response in the broader realm of healthy debate.

This book has grown out of research undertaken in the early 1980s towards a PhD in International Relations, at the University of Kent, Canterbury. The thesis itself has provided a framework which has been modified extensively and updated in the light of the continued growth of biotechnology. My thanks, therefore, go to Professor A.J.R. Groom and staff at Kent, and to all my colleagues who have given me much support and encouragement. Special thanks must also go to the administrators of the Recombinant DNA Collection at the

MIT Archives, in Boston. Finally, I must acknowledge the great help and unstinting encouragement given over the last few years by my wife.

Introduction

It is only by means of the sciences of life that the quality of life can be radically changed. The sciences of matter can be applied in such a way that they will destroy life or make the living of it impossibly complex or uncomfortable; but unless used as instruments by the biologists and psychologists, they can do nothing to modify the natural forms and expressions of life itself. The release of atomic energy marks a great revolution in human history, but not (unless we blow ourselves to bits and so put an end to history) the final and most searching revolution.[1]

Aldous Huxley made this observation in 1946 in a foreword to his book, *Brave New World*, first published in 1932. Of interest is his inference of a future biological revolution capable of radically altering the quality of life, not least through its modification. Huxley brought attention to hypothetical social, political and philosophical consequences of such a revolution. In the last decade some have described advances in genetic manipulation in terms akin to Huxley's forewarnings.[2] Hailed as something of a scientific revolution, the techniques of genetic manipulation heralded a very marked advancement in the ability of man to manipulate DNA, the controlling substance of all life. Whatever the long-term future holds for extrapolative developments of genetic manipulation, there will undoubtedly be philosophical, ethical, social and political questions to be addressed. Many of the claims relating to the interference with life through application of the new techniques were perhaps premature, although one characteristic linked them all. Such claims were primarily a consequence of the uncertainty that surrounded the earliest developments of *in vitro* procedures to manipulate DNA in the laboratory. In

1

addition, the very scientists at the forefront of interest in developing the new methods publicly, and somewhat dramatically, expressed fears of conjectured physical risks associated with the work.

In short, controversy raged in the mid-1970s following the publication of two letters in internationally respected science journals.[3] These called for thorough investigation of potential biohazards associated with the 'isolation and rejoining of segments of DNA' utilising a bacterium, strains of which were a natural inhabitant of the human intestinal tract. Could such experiments introduce into this indigenous bacterial parasite viral DNA capable of causing cancer in the human host? Could strains of microorganism acquire resistance to antibiotics previously missing? How would fragments of DNA 'behave' when isolated from their full strand and inserted into a bacterial host? Would there be long-term evolutionary consequences if manipulated organisms were released into the environment? What would happen if DNA from 'higher' and 'lower' organisms, were mixed? Could bacterial hosts 'express' DNA from higher organisms, including humans? Such questions fuelled broad discussion. Further, although rarely stated publicly at the time, some scientists were worried about the potential that the new techniques of genetic manipulation might have for the development of biological weapons.

Participants in the debates included scientists, non-scientist pressure groups, universities, the news media, government departments, specialist committees, various domestic and international organisations and even the European Community. By the end of the decade, the issues were being addressed within and across more than 30 states of the world. If scientists themselves displayed initial uncertainty, then concerned, if inadequately informed, non-scientists found themselves even more in the dark, to the extent that participation in forums of decision-making itself became a contested issue in many countries. Drama, sensationalism and attempts at 'rational' assessment were all intermixed in the, at times, heated exchanges. Guidelines for using the new techniques were developed, implemented, disseminated, compared, relaxed, but on the whole grudgingly accepted. Legislation loomed within and, in the case of Europe, over individual countries, although in the event receded. But such is the nature of our technological world that new ideas of intellectual or commercial promise will usually thrive despite adversity, bureaucracy and even risk. By the 1980s genetic manipulation had achieved the status of 'standard practice' in microbiological laboratories, and commercial products of the new revolution were being marketed. The patenting of new forms of microbiological life produced by genetic manipulation is acceptable, and indeed so is patenting of the laboratory procedures themselves. Databanks are being established for the deposit of decoded DNA and machines have been produced capable, although in a limited fashion, of synthesising strands of new DNA from component chemicals. If decoding and manipulating DNA represent the first steps, then writing new messages in the code may represent the next.

Exciting as these developments have been, their true industrial impact is being felt in combination with other more established microbiological, genetic and biochemical processes. The field as a whole has come to be termed 'biotechnology'. A recent OECD report, stressing the need for a standardised definition, observes biotechnology to be 'the application of scientific and engineering principles to the processing of materials by biological agents to provide goods and services'.[4] 'Scientific and engineering principles' are taken to cover a variety of disciplines, but in particular microbiology, biochemistry, genetics and biochemical and chemical engineering. 'Biological agents' refer to 'a wide range of biological catalysts but particularly to microorganisms, enzymes and animal and plant cells'. Similarly, 'materials' are taken in a broad sense to include both organic and inorganic components. The authors of the report go out of their way to avoid too narrow a view; for example, seeing biotechnology as essentially genetic manipulation; or too wide a view, including all activities involving living materials as biotechnology. Emphasis, in particular, is given to 'technology' being a scientific activity rather than an industry. The link with industry is in the application of the technology to produce 'goods and services' covering a variety of areas such as pharmaceuticals, biochemicals, foodstuffs, and services such as water purification and waste management.

A plethora of publications has appeared addressing issues surrounding genetic manipulation and more recently biotechnology, from journalistic accounts to works of some considerable academic depth.[5] Apart from book-length studies, extensive coverage has appeared in the numerous journals devoted to the relevant areas of science or science developments in general. The interested reader should have no difficulty in keeping abreast of progress in biotechnology through the pages of reputable science news journals. Yet, despite worldwide interest shown in developing biotechnological industries and basic research, worldwide dissemination of information, the worldwide potential impact of biotechnology success, and international competitive factors, there has been little or no attempt to examine the *international* characteristics of the case. This book will try to do this, but with a dual purpose in mind. On the one hand, it is hoped that those with a general interest in biotechnology will gain from the added international dimension. In taking an international orientation I shall draw upon the literature from the academic field of international relations, but in such a way that a second focus of this study will be to stress the importance of the development of biotechnology as a topic of relevance to international relations. However, a number of key issues will be addressed within these aims. Whether biotechnology is considerably more than simply genetic manipulation or not, there is no doubt that the controversy of the 1970s was fairly specific to the innovative microbiological techniques and practices involved in splicing DNA. In as much as the issue of risk or hazard was and is an international problem, then genetic manipulation will receive much attention in these pages. Drawing upon the history of the growth of widespread concern over conjectured hazards involved in manipulating DNA, I shall illustrate the difficulties entailed in applying rational risk assessment under conditions of uncertainty. These difficulties can in turn raise questions of participation in decision-making forums and, as far as the application of safeguards is concerned, lead to international disharmonies. In the long term all levels of society could be affected by biotechnology. At the centre of attention is the manipulation of the components of life itself in such diverse ways that hard choices may in future be required. Broad participation and openness will be essential. While for international relations, as a field of study, there are important issues that do not readily fit traditional viewpoints which tend to overemphasise the state as the most relevant political unit to study.

Biotechnology's many differing interested parties operate both within the boundaries of individual countries and across them. Organisations of a variety of forms communicate, compete and cooperate in an international environment. Aside from the questions of risk, the benefits to society and commercial gains to firms also display international characteristics. Indeed, biotechnology in the future is likely to increase in importance greatly within what can be termed the 'international political economy'. Economic, political and social factors are all involved as research and development pushes forward the frontiers of knowledge, creating new products and processes of production.

Agriculture, pharmaceuticals, medicine, chemicals and weaponry are all beneficiaries of the advances being made. The future has much in store in this respect. But we should not ignore the lessons of the past, albeit the recent past. Although genetic manipulation is only a foundation stone in the more extensive edifice of modern biotechnology, it was in the last decade the central focus of attention, and to many remains so. Despite a decline in the perception of direct hazard to human health, through inadvertent releases of novel microorganisms endowed with pathogenic qualities, we should not forget the lessons that can be learnt by examining the way the international community responded to the threat of unknown risks. As long as people perceived that there was a potential for untold damage, perhaps in the form of a carcinogenic virus, then the responses of decision-making bodies are important. The publication of calls for caution in the use of genetic-manipulation techniques by those most likely to want to use them was in itself dramatic, and the subsequent worldwide 'moratorium' on their application was without precedent in the history of science. Genetic manipulation provides a rare example of speculated hazards being addressed from the very discovery of the new techniques. It is simply fortuitous that the risks were revealed over time to be less than initially

speculated, as diversity in the genetic mechanisms of higher and lower species was revealed.

From the start, industry showed a keen interest in the development of the new science, and, as of 1984, 1200 firms worldwide were active in biotechnology.[6] In addition many academic establishments are pushing back the frontiers of bio-technology and interestingly, in many cases, there is consider-able cooperation between the two groups. Today, the first stage of the new biological revolution is being consolidated, especially as far as genetic manipulation is concerned. Zeal and excitement prevalent amongst those early pioneers of methods to rearrange DNA in the laboratory have now given way to a more sober period of refining the technology. Much older biotechnological techniques (even ancient as far as fermenting is concerned) are themselves benefiting either directly from additional inputs from genetic manipulation or from simply the general increase in awareness and interest in such things biological. The old and the new are finding them-selves to be mutually reinforcing.

Yet the start of the 'biotechnological revolution' is the focus of this book. The committees and organisations in the United States, the United Kingdom and other countries which put into operation safety guidelines will be examined alongside international conferences and international organisations, such as the European Science Foundation, the International Council of Scientific Unions and the European Community. Attention will be given to pressure groups and individuals important in the proffering of criticism about the decision-making procedures paying particular attention to the politics of the issue area. In all, the approach is to return to the dawn of genetic manipulation in order to trace the emergence inter-nationally of procedures and policy to cope with the possibility of potential hazard in the application of the new science, aware of the difficulties of decision-making under uncertainty and the political nature of such attempts. To assist this task an analytical framework is developed in the first chapter derived in part from the literature on organisational decision-making and operational 'systems'.

Any serious study should, however, make clear the underlying questions being addressed. It is all very well outlining the analytical framework, important as that is, but that framework has to be applied with some inquisitive end. Questions must also be selective as no one study can hope to cover everything that is deserving of our attention. The questions that have influenced this study have in part reflected the assumption that the early processes of decision-making, following the birth of the new biological techniques of genetic manipulation, have influenced the subsequent international rise to maturity of those techniques within biotechnology. The questions also derive from concern about the wider issue of technological assessment where uncertainty prevails, and opinions and objectives of those active in the debate are correspondingly diverse. There is, therefore, an unequivocal belief that there are lessons to be learnt.

To what extent were there political biases in the development of safety guidelines and their implementation? The history of the emergence of genetic manipulation has often been described as a controversy. This implies differing political attitudes in the sense of participants to that controversy holding conflicting interests or, at a more general level, conflicting values. From this it is but one step to ask how that politicisation of the issues affected the decision processes and decision outcomes domestically and internationally.

Was the range of options considered in the important centres of decision-making unnecessarily narrow? Given that decision-making is unlikely to be rational in the strict definition of that term, it nevertheless becomes appropriate to ask what range of alternatives was identified, on any serious basis, and, by inference, what was excluded. If choice structures were indeed limited, it is further necessary to ask why. In a wider sense we may ask whether the issue area itself was too narrowly defined as safety in the use of the new techniques. Were broader moral and philosophical issues left to the periphery?

Were safety controls searched for within existing operational frameworks on the basis of an assumption that the frameworks themselves should not have to accept more than minimal change? This question involves a defining of the boundaries of decision and organisational systems where participation and participants' roles are critically important. Not least this raises the further question of whether or not there is any evidence of any groups engaging in activities associated with 'role defence'. *Was there any concerted effort to try to apply the lessons of other technologies regarding safety issues and legitimacy in the development of policy?* In other words, was there a tendency to see the issue of genetic manipulation in isolation? Much might have been learnt, as much in future might be learnt from genetic manipulation, by examining other technological phenomena that perhaps displayed similar general characteristics, notably of low probability, high potential consequence risk. Were there any improvements over the way that risk was handled compared with the experiences of other technologies?

How effective were communications between the many national, transnational and international groups? If uncertainty was a primary feature of the early days of the new biotechnology, then the efficacy of information dissemination could be seen as very important. Associated with this is the need to identify that information which is particularly significant to assessment of hazards.

Were there any particular problems at the international level? Given the lack of any international source of central authority, were the issues prevalent at the domestic level further complicated by international and transnational differences? Who participated in international interactions and what interests and values did they represent? If further complexity was apparent, have the international problems been at all resolved?

1 Biotechnology and the New Microbiology

Manufacturing products useful to man, including the provision of a variety of alcoholic beverages, by applying biological processes has a history which can be traced back to antiquity. Nevertheless, biotechnology enjoyed something of a rejuvenation in the last decade, if not a revolutionary thrust forward, with the advent of certain discoveries and technical laboratory breakthroughs. In particular, the technical skills associated with what has been called 'genetic manipulation', or more evocatively 'genetic engineering', are prominent. To stress the point already raised, genetic manipulation is by no means the only important element comprising biotechnology. Yet the early 1970s brought tremendous interest in the new techniques to the extent that two leading individuals, central to the controversy surrounding genetic manipulation, in retrospect observed:

but with the new recombinant DNA tricks the genetic engineering of micro-organisms, and later of higher plants and animals, would help to shape the world of the future. Without doubt molecular geneticists now had the power to alter life on a scale never before thought possible by serious scientists.[1]

There is no doubt that the overall promise of biotechnology, in reflection of the breadth of the component science, is diverse. Yet biotechnology should be seen as a collective field given the interrelatedness of many of the scientific techniques and the knowledge underlying it, and the common but fundamental biological nature of it. However, the component parts in themselves may provide great interest. The modern history of the impact of genetic manipulation on microbiological research techniques and the corresponding attention *à propos* its safety has to many commentators represented an issue area. Of all the developments which underlie the rapid growth in biotechnology, none has demonstrably been so controversial. Much can be learned by returning to the dawn of the technical innovations comprising genetic manipulation to examine the case study from an international perspective. Indeed, even the commonly raised question of safety in the research, given perceptions of the time, can tell us much when addressed anew.

Moreover there is a case, to be supported, that the traditional boundaries between certain fields of study which have historically developed are rather arbitrary. International relations as a relatively recent field[2] has borrowed more than most in an interdisciplinary fashion. Economics, politics and sociology, to name but a few, have retained their separate identities to the extent of neglecting many overlaps. These overlaps are at both the domestic and international level.[3] Technology can represent an important issue to all of these fields, and an international focus, in particular, should emphasise this. As a case study, the impact of genetic manipulation on society is relevant for its economic importance, its wider social costs and benefits, and the overtly politicised scrutiny it received. Potentially new and existing products and production processes have brought interest from established and new specialist firms. Society is likely to benefit in the long term from new knowledge about certain diseases caused by both hereditary factors and failures in the genetic ordering of biological functions in people and animals. Yet society may also have to face ethical and moral questions, not unlike those surrounding the achievements of 'test-tube baby' research and the experimentation on human embryos. Such questioning has already begun because of the potential in the longer term for 'interference' in the human genetic structure or for the aborting of babies with genetic defects, when bio-technology assists their identification. Safety issues have in their turn raised overtly political issues regarding acceptable risk, and participation in decision-making where constraints on the research were felt necessary. But superimposed on all questions of this sort is the international nature of our world which defies cultural, economic, social and political harmony not just within societies but also between them.

It becomes necessary, before proceeding to outline the framework of analysis to be applied in the study, to say something of the laboratory techniques which comprise genetic manipulation.

GENETIC MANIPULATION AND INTERNATIONAL RELATIONS: A TRANSNATIONAL EXPERIENCE

The Science of Manipulating Life

An important point to bear in mind is that this discussion of the techniques of genetic manipulation is assisted by the benefits of hindsight. The next chapter will indicate the origins of concern in an historical context and show how fears developed before some of the methods presented here were achieved. Indeed, the general decline of expressed fears, which became evident since the late 1970s, was related in part to some limitations of the techniques themselves, further influenced by attention being drawn to natural processes which appeared to produce results similar to the genetic manipulations. It should also be said that the term 'genetic manipulation' is not standardised within the scientific community and amongst non-scientist commentators. A common alternative is to talk of 'recombinant DNA' techniques which, although seemingly technical to the layman, more accurately fits with the scientists' specialised jargon. Generally speaking, differences in definition reflect two issues: first, the nature of the phenomenon to be categorised, which will be briefly described below; and secondly, the label applied to this phenomenon which, as already mentioned, may be quite evocative.[4]

It will not be necessary to enter into great detail in describing the science behind the ability to manipulate DNA in the laboratory. What is more important is to give some impression of the innovatory significance of the new techniques. That significance lies in the biological function of DNA. When Charles Darwin presented his theory of evolution it had been formulated without knowledge of the mechanisms of inheritance. Yet it is the modern understanding of genetic structure which underlies an explanation of the process of evolution which he advanced. Genes have become identified as the determinants of hereditary characteristics of all living organisms and are associated with a particular kind of molecule known as deoxyribonucleic acid, DNA for short. The importance of DNA is emphasised in that the principal 'dogma' held by biologists today is that DNA makes RNA (ribonucleic acid, discussed below), RNA makes proteins, and proteins make everything else.

In 1869 Fredrich Miescher, a Swiss biochemist, isolated a substance given the name 'nuclein'. He later showed that sperm nuclei consisted of approximately 60 per cent nucleic acid and 35 per cent protein-like compounds. However, until the 1920s, the prevalent view was that biological specificity resided in proteins and not nucleic acid, the proteins being made up of amino acids. It was shown that further purification of DNA improved the efficiency of transferring traits, and later in 1952 that when a virus infected a cell, only DNA entered. Thus DNA became identified as the main determinant of genetic information.[5]

Identification of DNA as the main genetic determinant was not the same as understanding its chemical structure and the means by which it functioned in this role. The most important insight into the processes of heredity, and indeed evolution, was to come from the work of James Watson and Francis Crick, published in 1953. In effect they *deduced* the precise three-dimensional structure from the available data, which they then used to propose a completely new suggestion as to how genes could replicate (and how mutations could arise).[6] Part of the existing evidence that Watson and Crick were to use was that experimental tests showed that four of the known components of DNA, termed *bases*, appeared in quantities suggesting some form of pairing. Other researchers were to suggest that two of the bases, *adenine* (A) and *thymine* (T), occurred in approximately equal amounts, as did the other two, *guanine* (G) and *cytosine* (C). Further, adenine and guanine were of one type of molecular base known as a *purine*, while thymine and cytosine were of another type known as a *pyrimidine*. The pairings therefore included one purine with

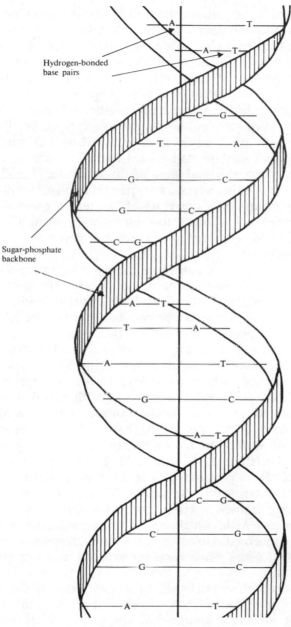

1.1 The structure of DNA

one pyrimidine, but always the same ones. Although evidence had shown that the paired bases could differ quantitatively in samples, a factor which assisted the identification of the above pairings, it was also evident that the total quantities of purines and pyrimidines were apparently equal. Incorporating further information gained from X-ray photographs of crystalline DNA, Watson and Crick pieced together a model of DNA that has subsequently enjoyed universal acceptance. With A always paired with T and C always paired with G, the DNA molecule was found to comprise two helices which rotated around the same axis and were of the same size. Finally, the chemistry of DNA, which was partly known, involved a phosphate group and a sugar, which in the Crick and Watson model formed a 'backbone' to each helix, between which the paired bases were arranged. Figure 1.1 illustrates the now famous structure.

It was the pattern of pairing in the bases that turned out to be the key to both the transmission of genetic information and the process of DNA duplication. Each of the individual bases would be linked to one of the two backbones. The base, the sugar and the phosphate group at that point was termed a *nucleotide* and it was the sequence of such nucleotides, which could be enormously long, that, according to Crick in 1958, provided the code which could lead to the production of at least 20 amino acids. For the purposes of this book it will not be necessary to refer to nucleotides in any way more detailed than in terms of the associated bases. Thus in describing sequences of nucleotides for simplicity we can merely identify chains of the letters A, C, T and G. In any case, this really gives a better impression of the idea of a genetic 'code', identifiable in terms of these letters. We use the term 'gene' to identify long sequences within the DNA chain perhaps 1500 base pairs in length, which code for particular traits in the organism concerned. However, a simple bacterium might have around 3000 genes while each human cell contains perhaps 3–4 million.

Sequences of three nucleotides (or three letters of the genetic alphabet) were found to be responsible in turn for the sequencing of the amino acids, chains of which structure proteins. However, to cast further light on the method by which this occurs it becomes necessary to introduce the role of *ribonucleic acid* (RNA), a single-stranded derivative of DNA. Chromosomes which carry genes are found within the nucleus of living cells, while the synthesis of proteins occurs in the surrounding cytoplasm. It is a special type of RNA (known as messenger RNA) which carries the genetic message out of the nucleus. RNA can form under special conditions when the DNA double helix separates and one of the two strands becomes a template temporarily attracting free bases.[7] A strand of RNA forms which is complementary to the original DNA strand acting as the template, and is effectively identical to the displaced DNA strand. The messenger RNA, when formed, separates from the DNA template and leaves the nucleus carrying with it the sequences of three bases which determine the subsequent ordering of the amino acid chains.[8] Thus the original genetic code within the DNA strand is *translated* into chains of amino acids or proteins.

However, the DNA code is not only responsible for the transmission of information within an organism, it is also responsible for the passing of information to subsequent generations of that organism. Under appropriate chemical conditions the bonds of the paired bases (A and T, and C and G) weaken, and the double helix unwinds and separates. In the presence of suitable enzymes (proteins that act as chemical catalysts) and freely available nucleotides of the four kinds, a new strand will form by complementary bonding to the exposed nucleotides of each of the older, but now separating, strands (see Figure 1.2). All the genetic information of the original DNA double helix is incorporated in each of the resulting copies. At cell division the appropriate conditions for this to occur exist, its importance summarised by Clifford Grobstein:

At the level of molecules, like begets like through DNA replication. This phenomenon occurs in the reproduction of every organism on earth and it has been happening, so far as we know, ever since the first time life emerged eons ago. As a chemical process, the doubling of DNA is undoubtedly the most prolific and portentous of nature's entire bag of tricks.[9]

Not all replication is perfect, however, and variation enters the endless production of copies through mutation. One means

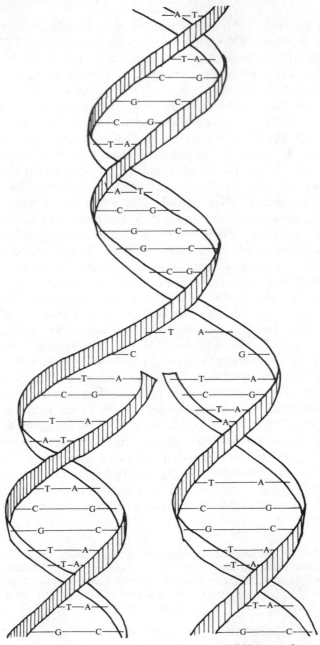

1.2 The separation and replication of DNA strands

by which mutation can occur is if in DNA replication a base pair becomes deleted or two bases are mismatched. Future generations would then be affected, although some correction would occur in the case of a mismatched pair on DNA replication, where the originally correct base in the pair would acquire, on cell division, the correct complementary base. The incorrect base would, however, acquire its complementary base ensuring that all future replications in its line would carry a base pair now fully substituted for the original pair at that point in the sequence. A new base pair could alter the RNA message carried out of the nucleus and in turn alter an amino acid in a protein chain. I should, though, point out that mutation is complicated by the existence of mechanisms which can correct such errors, and therefore any major species change for example would involve many errors overcoming a number of biological obstacles to become established permanently and in serendipitous combinations.

To a large extent the unravelling of DNA and its biochemical functioning, with all the attendant mechanisms, has given a model which can accommodate Darwin's much earlier theory of natural selection. However, the 1970s were to see major breakthroughs in the ability of man to *manipulate* DNA and increase his understanding of its functioning, through the application of biochemical laboratory techniques.

For centuries man has used selective breeding techniques to develop 'errors' that have occurred in breeding populations, through mutation, when resulting characteristics were seen as desirable. Thus agricultural and pet animal selection have taken place. Natural selection in the Darwinian sense also makes use of these mutations, but the success of both forms of selection is dependent on the random occurrence of desirable mutations. The importance of recombinant DNA techniques is that for the first time genetic change can itself be directed by manipulation of nucleotide sequences, including the transfer of known sequences from one strain or species into another. The techniques themselves followed a number of discoveries made in the 1960s and early 1970s, relating to the recognition that not only were proteins a product of genetic translation but that certain proteins were themselves essential for these genetic processes. A major function of proteins is to act as catalysts, or

1.3 A sequence of DNA recognised by restriction enzyme endonuclease Eco RI.

enzymes as they are known, and it was certain types of enzymes that provided one of the most important means to manipulate DNA. Work in 1962 on a particular bacterium that was to become important to genetic engineers showed that a bacterial virus which grew on one strain of the bacterium grew poorly on another strain. It seemed that the bacterium, *Escherichia coli* (or *E. coli* for short), had a mechanism by which it could protect itself from foreign DNA. It was found that a particular enzyme, called *restriction endonuclease*, introduced a number of breaks in the backbone of the infecting viral DNA, but did not cut the DNA of its own cell due to a further set of enzymes in the cell which provided protection. In effect, the code embedded in the long sequence of DNA becomes corrupted when the sequence is broken. It was the abilities of such restriction enzymes that were to provide a powerful, although conceptually simple, tool to enable the direct manipulation of DNA derived from any organism.

Critically important to the breakthrough that has borne the genetic manipulation revolution was the discovery that restriction enzymes cut DNA on the recognition of particular sequences of nucleotides. Moreover, in 1972, it was found that a restriction enzyme from *E. coli*, known as endonuclease *Eco* RI, could cut the double strand of DNA on recognition of a

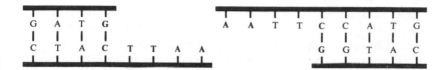

1.4 DNA spliced by endonuclease Eco RI.

sequence of six nucleotides, or paired bases, but leaving – and
this was the beauty of it–a cut across the two strands both
staggered and identical. Using the fairly common diagrammatic way of representing paired bases (rather like a ladder
with the nucleotides represented by the rungs and the
backbone by the sides) the sequence identified is shown in
Figure 1.3. The enzyme then makes a staggered cut between
the adjacent A and G on each backbone.[10] As a result each
now separated section of DNA ends with a short 'tail' of
single-stranded DNA, which, because of the symmetry of the
cut, is both identical for each section and capable of bonding
with any other section of DNA cut by the same enzyme, as
shown in Figure 1.4.

Often called 'sticky ends', sections of DNA cut in this
fashion will always be able to join together even if the source
of the DNA is from completely different organisms, although
to complete the bond the two backbones have to be rejoined
with further chemical treatment. Identification of a variety of
restriction enzymes has since ensued, each recognising
particular sequences of bases.[11] Overall, this method of
manipulating DNA overcomes a number of difficulties. Not
least it provides a means both to break strands of DNA and to
enable their joining.[12]

Cleaving and joining DNA is, however, only a part of the

whole process involved in obtaining something *useful* from recombinant DNA techniques. Further modifications need to be outlined. *Eco* RI, for example, cuts DNA on average every 4000 base pairs. Taking DNA from two sources and mixing them could lead to a number of possibilities regarding the hundreds of thousands of likely pieces. All of these fragments would have exactly the same sticky end enabling a fragment to circularise as its own ends join, or each fragment could interact with essentially equal probability with any of the fragments from either source, or it could join with DNA from its own organism. When two fragments have joined they in turn could react like a single piece, with the same possibilities. Thus the potential linkages are very complex. Because of the crucial importance of the organisation of chromosomes in order for the regulation and functional *expression* of genes to occur, there is very little likelihood of anything viable as an organism resulting from such a mixture of DNA.

In order to get something useful from the mixing of genes from different organisms certain simplifications assist. By avoiding complex mixtures of DNA – for example, as one author has put it, using DNA from ducks and oranges[13] – more carefully designed combinations could engender expression. A powerful modification to the experimental procedures involved in splicing DNA is to make sure that one of the sources of DNA has the power of self-replication. Replication of DNA requires the DNA to be physically linked to special sequences of DNA which code for replication functions. These particular genes are relatively rare, and bacteria which contain about 3000–5000 genes only contain a single set of replication genes. Nevertheless, a particularly useful means of gaining access to replication functions was developed, through the use of *plasmids*. These are small circular pieces of DNA which are found in some individual bacterial cells, independent of the cell's large circular chromosome, and are capable of self-replication. Almost every known form of bacterial cell can house plasmids, although only a very few individual cells will actually contain them.

In 1973, Stanley Cohen and Herbert Boyer led an experiment using the plasmid pSC101 (plasmid Stanley Cohen 101), selected because of its property of having only one *Eco*

RI recognition site within its circle of DNA. Thus when cut, the plasmid could link with foreign DNA also cut with the same restriction enzyme such that a new hybrid circular plasmid could form. Cohen and Boyer used 'foreign' DNA derived from another plasmid in this first experiment, but the way was opened for similar experiments using DNA from other sources.[14] However, not all cells will contain the independently replicating plasmids, and some means of selecting which cells actually contain the plasmids becomes necessary.

Identification of cells containing plasmids has been facilitated by the fact that many plasmids contain one or more genes which confer resistance on the cell to one or more antibodies that would normally kill it. These genes specify enzymes that can break down the drugs enabling experimenters, who introduce genetically-modified plasmids into plasmid-free cells, to identify future generations of cells containing plasmids by introducing the resisted antibody. Flexibility exists in the approach in that different plasmids can be used which respond to different antibodies. But the identification of cells containing plasmids is only part of the problem. It is also necessary to be able to tell which plasmids initially acquired the *additional* DNA after genetic manipulation and which did not. In order to do this, use can be made of plasmids that contain genes which code for resistance to more than one antibody. The plasmid must then be cleaved by a restriction enzyme that recognises a nucleotide sequence that occurs within one of the gene sections coding, for example, for one of two resistances. Insertion of foreign DNA at such a point inactivates that resistance by altering the necessary sequence of nucleotides. Resulting cells displaying resistance to only one of the two antibodies most likely contain foreign DNA.

Other methods of screening exist which enable the experimenter to identify cells which contain the modified plasmids, making the use of plasmids, as a means to replicate genes of interest, a powerful tool. But plasmids are not the only way to acquire replication. A second method is to use certain classes of viruses known as *bacteriophages*, or *phages* for short, which grow on bacterial cells. In bacteria phages can be incorporated into the chromosomal DNA itself and some phages were to display particularly interesting features. A phage often used was *lambda phage*, which even with sections of its DNA deleted could still replicate. In 1974, researchers took advantage of its properties to generate special forms of lambda phage into which foreign DNA could be inserted, but which had only a limited number of restriction enzyme cleavage sites and a large deletion of DNA. In addition, the phage would not propagate in *E. coli* without the foreign inserts. Thus the foreign genetic inserts could be carried in the phage as it infected bacterial cells and as the cells replicated.

By way of terminology the plasmids and phages used in replicating inserted DNA are collectively known as *vectors*. In effect they allow the magnification of quantities of DNA, once the foreign DNA is inserted. However the foreign DNA and vector combination requires the services of a bacterial cell in which the replication process proceeds. This is termed the *host* and it is common to talk of the host and vector taken together as a *host–vector system*, in the context of the manipulation of DNA. They are standard microbiological terms. Nevertheless, we cannot rest here in terms of our simplified description of genetic manipulation. It is one thing to acquire large quantities of particular DNA sequences through insertion in a host–vector system, it is quite another to obtain the final product for which that DNA codes in its natural biological location.

As work proceeded in the 1970s, it became apparent that the problem of functional expression had two sides to it. Attention was first directed to the task of getting the DNA from one lower organism to express in another lower organism, or *prokaryote*. The second task was the expression of DNA from higher organisms, or *eukaryotes* whose cells contain a nucleus, within lower prokaryote hosts. Early successes in obtaining the expression of foreign prokaryotic genes in a prokaryotic host led researchers to believe in general that this would be the result when prokaryotes and lower eukaryotes were the source of the donor DNA. On the other hand, the expression of DNA from higher eukaryotes when inserted into bacterial hosts proved to be a much more complex problem.[15]

Let us look again at the requirements for successful

expression. Initially the DNA has to act as a template for the formation of messenger RNA, which must then carry the genetic message to be 'translated' for the production of proteins. But the correct execution of each of the involved steps depends on the correct recognition and interpretation of certain 'start' and 'stop' signals embedded in the DNA code. *E. coli* has proved to be a powerful host in that it can 'read' these signals in the DNA from other prokaryotes, although in addition expression would also require the translation of the code. Whatever difficulties exist with prokaryotic DNA they are considerably less than those associated with eukaryotic systems. It was found that in nearly all mammalian and vertebrate genes, and in the genes of eukaryotic micro-organisms such as yeast (although less frequently) there were interspersed inserts of non-coding DNA, termed *introns*. Simplified, it has been shown that the DNA template initially allows an RNA transcript to form which subsequently has the intron sections 'spliced' out. It is the modified RNA which provides the code for the ordering of amino acid chains, and thus proteins. It is probably the lack of appropriate 'splicing' mechanisms in prokaryotes that represents the most important barrier to the expression of foreign genes taken from eukaryotes.

For the scientist there are means by which some of the problems can be overcome. For example, adequate signals may be added to the foreign gene prior to its insertion in the vector used, thus adapting the foreign DNA to the requirements of the host, for gene expression. A tendency for *E. coli* to degrade proteins formed from the translation of foreign DNA (in a fashion using its natural self-protecting mechanism for removing proteins derived from internal mutations) has been overcome in the laboratory by attaching the foreign DNA to a particular host gene. All of this, however, adds to the need for careful experimental design if anything useful is to be derived from foreign DNA placed in hosts which provide replication functions. In the next chapter I shall show how ignorance of some of these difficulties led to fears of unknown consequences if eukaryotic DNA was to be randomly inserted into prokaryotic hosts.

Many of the problems stem from the fact that the methods

of 'splicing' DNA do not separate discrete genes. If an average gene contains some 1500 base pairs, and the commonly used restriction enzyme *Eco* RI cuts DNA approximately every 4000 base pairs, it can be seen that the resulting fragments will contain some partially included genes. Perhaps the most profound method of overcoming the problems faced in attempting to obtain the expression of useful genes in host organisms is to attempt to obtain the gene in question in isolation. It has already been said that double-stranded DNA produces single-stranded RNA. However, in certain cells the quantity of RNA produced in real terms, derived from a particular gene, may be far greater than the original proportion of DNA, comprising the gene, relative to the total DNA in the organism. In some cells the RNA from a particular gene may constitute as much as 30 per cent of the total RNA in that cell. In such quantities it can be isolated and purified by conventional biochemistry techniques. The experimenter has found that this can be utilised by inducing in the laboratory a second strand of bases, attached to the single-stranded RNA. Subsequently this can be made to separate, enabling the new strand to operate in the same fashion as separating DNA, in turn attracting its complementary second strand. Thus from purified RNA the experimenter can work backwards to create the best part of the original double-stranded DNA, which in turn can then be inserted into host–vector systems.

Again working backwards it is possible to synthesise a strand of DNA from a *known* sequence of amino acids. New methods in chemical synthesis coupled with the application of enzymes enable the construction of any given nucleotide sequence. Indeed, such have been the advances that computer-guided machines can undertake the repetitive chemical procedures, while other machines can decode particular DNA sequences. In effect we are developing skills in the 'reading' and 'writing' of the DNA code. Applying these skills has led to the sequencing of a variety of genes from different organisms enabling the establishment of data-base 'libraries' of decoded genes.[16]

The 1970s may have seen the tremendous growth of interest in developing and using the new techniques of genetic manipulation, but in the 1980s these are intermixed with other new experimental techniques[17] and traditional methods, within the wider field of biotechnology. In the following chapters much will be said about the controversy which arose over the safety of genetic manipulation. For the most part that controversy may have focused upon safety, but questions also arose over the definition of genetic manipulation and the applicability of safety restraints on experiments which might produce the same end product, without using the newer techniques. Added to this was the growing awareness that variations on inter-species genetic recombination occurred in nature itself.[18] Consequently, the issues were quite complex and were characterised by much uncertainty about the hazards posed in using the above described methods, further discussed in Chapter 2. As science has an international tradition, it would not be at all surprising that the debate crossed national boundaries. From the beginning the interest in the potential for genetic manipulation was clearly international, but so too was the early expression of concern.

Towards a Framework of Analysis
Any framework that we apply to the study of the international features of the case of biotechnology must take account of certain prerequisites which come to mind. We must, for example, be able to include analysis of the activities of a variety of different types of groups or organisations, including the pattern of their interactions whether they are domestic, transnational or international. It is clear that scientific organisations and committees, working parties, government bodies, private industry, universities, pressure groups and international organisations were all of importance in the unfolding story of the emergence of the new biotechnology. Accordingly, our framework must be quite flexible. In examining transnational links between such groups, there is a need to map the pattern of their communications, identifying key points of exchange of information and ideas, viewpoints and policy, and establishing the dominant information routes. This will also necessitate an examination of the conflictual elements of their interaction as well as the cooperative. Taken with the identification of the key actors on our transnational

stage, the tracing of linkages between them helps facilitate the recognition of the overall political structure of relevance to the issue area. Although the development of responses to the early perceptions of hazard was international in scope, it should be said that the emergence of policy in both the United States and the United Kingdom was particularly noteworthy. Quick off the mark, these countries provided a significant lead which other states and organisations were both to monitor and to follow. The formulation of their policy and its impact will, therefore, be traced. Thus the study will embrace certain domestic interactions, but with an awareness of their international and transnational significance.

In the event, many states developed policy towards the conjectured hazards of genetic manipulation, which often involved the establishment of working parties or central committees charged with monitoring future developments and making policy recommendations to the appropriate government department. Yet certain facets of the procedures were common to all the important organisations faced with such tasks. Not least the decision-making forums were faced, at a minimum, with many characteristics of what J.D. Steinbruner has termed 'complex decision problems', encompassing the following features:

1. (a) Two or more values are affected by the decision.
   (b) There is a trade-off relationship between the values, such that a greater return to one can be obtained only at a loss to the other.
2. There is uncertainty (i.e. imperfect correspondence between information and the environment).
3. The power to make the decision is dispersed over a number of individual actors and/or organizational units.[19]

Taking these characteristics in turn, it could be said that the first feature applies to any political situation by definition, in that differing values are at the heart of all politics. The qualification that a trade-off relationship exists between values is less straightforward. In making decisions or choices there is always the opportunity-cost factor if resources are required to make and implement them. In these terms there is an obvious trade-off. That complex decisions involve trade-offs in values affected is less clear, especially in such 'zero-sum' terms as a gain to one is at the cost of a loss to another. In many instances this may be the case, including some decision-making which occurred over genetic manipulation, but in others the conflict of values itself may be as a result of misperception. The fear that a certain action might bear upon an individual's or group's values or goals could initiate an aggressive response, even if the choice to be made, in reality, would not affect them. Access to information, and knowledge of the respective values held by each involved group, might, for example, reduce such consequences of misperception. Suspicion and mistrust can promote value-conflict, while wider participation and input into decisions might help redefine apparent zero-sum choices as positive-sum, where all sides could benefit. Legitimacy is central to this. Decisions are more likely to be accepted if the process of choice making is seen, by all affected by it, as legitimate.[20] In the case of the choices surrounding genetic manipulation in the 1970s, it was apparent that some influential decision-makers wished to treat the problem in a way amenable to 'rational' assessment. On the whole, the tendency was to frame such attempts (despite the levels of uncertainty that prevailed) as a problem of setting risks against 'containment' or safeguard precautions. In general, the questioning of benefits arising from the new research was very limited. Rather it was the case that great benefits were assumed, the only real problem being one of time or complexity in their achievement. This was compounded by many stressing the tenets of academic freedom in investigation, to support continuation of research, whatever the direct benefits. Trade-offs did characterise the case of genetic manipulation in the above sense. Work could go ahead but with restrictions on the type of laboratory, while others, on the periphery of decision-making, questioned the benefits to be gained and the sanctity of freedom of research.

However, perceptions changed, although not necessarily reducing conflicts of opinion, but transforming the disputed central issues. Scientists, for example, who had advocated caution changed their views; while other groups, once voluntary guidelines emerged, began to call for legislation.

More recently, on the basis of increased knowledge, risks have been shown to be less than first thought. Underlying much of the evolving debate have been the mounting applications of genetic manipulation as it enters the toolbox of industrial bio-technologists. The politics of the safety issue have gradually transformed to encompass a greater input from industrial users of the technology as they embark upon environmental release experiments, test new products and realise commercial returns from applications of genetic manipulation. New safety issues have arisen as has greater competition, both between individual firms and nationally sponsored biotechnology development programmes. It will be necessary to investigate any new trade-offs between safety, industrial returns and differences in national policies towards acceptable levels of risk.

Uncertainty, the second feature of complex decision problems, was undeniably evident with the advent of the new biology of genetic manipulation. Indeed, it was uncertainty itself which led to the emergence of genetic manipulation as an issue area in the first place. Although in the assessment of technological risks uncertainty is often allowed for by the allo-cation of probabilities, in genetic manipulation there was insufficient knowledge upon which to found such estimates. Probability estimates also tend to reflect fairly structured problems, not characteristic here, with many diverse views held by the numerous actors. Much of the decision-making was in any case centred on the initial steps of how best to proceed, and how best to establish organisational frameworks to cope with further issues as they arose. Thus, uncertainty was of such a wide nature that value judgements were inescapable, given that differing priorities and goals emerged. More formally, the concept of uncertainty in this case included both *objective* uncertainty, where empirical evidence and scientific understanding were incomplete, and *subjective* uncertainty, where considerable differences in value judgements existed. Many scientists nevertheless extolled the virtues of a 'rational' approach.

I have already argued that the case of biotechnology, and in particular the controversy surrounding genetic manipulation, involves many organisations, groups, individuals, and firms.

Choices were not made, in terms of safeguards and their implementation, by single organisations (although some were obviously of more importance). If conflicting values, conflicting interests and uncertainty were evident, then these were reinforced by not just the range of actors which were directly involved, but also by the number of actors which saw one of the central values at stake as being the right to participate in decision taking and implementation.

Any framework we apply to the history of the growth of biotechnology in the 1970s must at a minimum take account of these transnational features of decision-making. A complex network of linkages existed which was faced with the features of uncertain choices. But more perhaps needs to be said about the nature of decison-making and organisations before moving towards describing the case of biotechnology.

The processes involved in the taking of decisions have been given very broad attention in the differing contexts of many academic fields. Economics, sociology, psychology and social psychology, politics and international relations all find the need to consider the making of decisions. Yet much of the decision-making literature is a response to dissatisfaction with approaches which emphasise the rational nature of decisions. Given the explicit appeal to 'rationality' made by many scientists concerned about the holding up of work while safeguards were debated, we should consider why their petition concealed a conceptual minefield. The problem of rationality concerns both the assumptions made about how individuals or groups actually make decisions, and how the analyst examines decision processes. These are obviously related, but nevertheless each reflects particular dimensions to the overall quandary. Perhaps the two-sided element can be summarised best by making the point that when decision processes are analysed it is easy to argue that if rationality is *assumed*, then analysis of procedures involved becomes more straightforward. In support of this there is the subtle rider that even if real-life decision examples do not reflect 'rational' activity, then they *ought* to. The normative side is often used, however, without an adequate appraisal of what rational decision-making implies in operational terms. Conceptions of rationality, even if imperfect conceptions, were most evidently

a vital component of the assessment debate regarding the risks associated with genetic manipulation. In particular, part of the politicised debate centred on whether or not more rational assessments should or *could* provide a base for decisions on the development and application of appropriate controls. Many scientists called for 'rationality' and 'rational assessment' on the grounds that this was the scientific way to proceed. Part of the reason for this was apprehension about unsubstantiated fears of risk, general politicisation of the issues, and even threatened legislation. The difficulties lay in that truly rational decision-making in any meaningful way is extremely difficult in some choice situations, and that imperfections in part of the procedures used can undermine the whole effort.

What then is a rational decision? Economists and others have long used the image of 'rational man' to investigate the way consumers or decision-makers make choices, where rationality can be taken to mean consistent value-maximising choices within specific constraints.[21] The critical features of the rational decision are the ability of the decision-maker to identify goals and objectives, which in decision groups implies the agreement of the members, and to identify the possible consequences of their decisions and the likelihood of their outcome. Then they must be able to bring both sides of the problem together through implementation, in order to maximise their gains.[22] All this is fine when the necessary information is available or calculable, which may be the case in certain constrained decision situations. But when the problem of uncertainty raises its head then rational assessments become extremely difficult, a problem compounded if the parties to the decision-making hold differing goals, interests or values.

Analytical models based on assumptions of rationality are seductive because large numbers of empirical facts can be interpreted by the application of a few simple assumptions about the goals that decision-makers are trying to achieve. Lacking precise information of the diary of events and the considerations of decision participants, the analyst, by assuming goals and rational action, can infer retrospectively the alternatives considered and attempt to explain rationally the course chosen. A rational choice by a group of many individuals, or a single person, would be precisely the same given that they have the same goals. Personification of organisations or even states is therefore acceptable if decision-making is seen as comprehensively rational, or as near that as practicable. In the case of genetic manipulation many different organisations and interest groups were active in the processes by which key decisions were taken. Activities in different states and in different organisations were linked, but these decision-systems could not be expected to agree on a single set of rational decisions. Compounded by the very uncertainty surrounding the case, value difference and differences in decision procedures led to a lack of harmony in international choices. Conflicts of interests and values between various actors were reflected in different goals, which would in any case produce differing outcomes.

Underlying the different interests held by scientists, specialist interest groups, industrialists, pressure groups and even governments or their administrative organs, were varied belief systems and values. Values, or sets of beliefs, both influence and are influenced by perception. Cognition and beliefs not least may structure the decision frameworks and the question at issue. The problem of perception, however, could take us into the realms of psychology and social psychology, embracing aspects of individual and group behaviour. Interesting as this might be, all we can really consider here is the assessment of actors' standpoints as they were revealed, especially with emphasis on conflicting interests and actor participation.[23] Thus, calls for rationality may have been heard but the competing interests involved made the decision processes inherently political.

Rationality may be a problematic concept, but it is not completely without value in considering decision-making in genetic manipulation. If its limits are recognised it still provides a useful component when the wider elements of decision-making are addressed. Not least we can focus on choice making in relation to limitations or incompleteness of information inputs, uncertainty of choice alternatives, uncertainty regarding consequences, and less than optimal outputs.

The problem faced is quite difficult in analytical terms. On the one hand, we need a means to address the processes involved in the taking of complex decisions appropriate to the specific case of genetic manipulation. This is necessary in that very real decisions were taken, as reflected in the publication of safeguard guidelines in many countries, amid some controversy. On the other hand, the groups involved in the decision processes were quite varied and active within and across the many different countries. Usually, approaches to the analysis of controversial technology tend to be geared up to the particular conditions of one state, or alternatively focus on intergovernmental relations. The need here is for a framework which can incorporate the different levels of analysis, from single organisation, through national, transnational, transgovernmental, to intergovernmental levels, in such a way that analysis can be directed within levels and across them.[24] Within the literature of international relations there is no definitive approach to such tasks. Conceptualisation of transnational and transgovernmental relations has become an acceptable part of the literature, but does not take us greatly beyond a means of defining particular sorts of actors and of including a wider range of issues than if we concentrated on states as discrete entities. Each issue area will bring, with it, its own particular pattern of interaction amongst those actors. Essentially the focus is on pluralist conceptions of political processes, where political decision-points are dispersed transnationally.[25] As far as this study is concerned, such concepts represent an appropriate starting-point. But in order to combine the dual elements of varied and dispersed decision forums with the need to examine the decision processes themselves, then we can usefully turn to the interdisciplinary offerings of *systems theory* and *theory of organisations*.

John Burton, an international relations scholar, takes an approach which is both distinctive and highly interdisciplinary. To Burton, systems provide an analytical device by which to break down many types of complex problems. In seeking explanations at the 'level of the whole', Burton notes that 'systems and sub-systems are wholes in themselves, acting within their environments of other systems and sub-systems'.[26] In order to identify the limits of appropriate analytical systems, which can overlap through different levels, Burton suggests we define each system in relation to the 'roles' fulfilled by the component units. Systems, therefore, comprise units of the same 'set', where sets can be identified by collective roles.[27] To elucidate, individuals or groups can be members of different sets, and hence systems, depending on the role they play in each. Scientists for example, are members of the international science community, which could be described in systems terms, but some might be involved in science policy decisions within government, and nearly all could be located in terms of membership of specialist groups reflecting their field or specialisation. In this sense, systems are 'open' and receive inputs and produce outputs in relation to their surrounding environment, or other systems. For conceptual purposes social and political systems have limits or boundaries, although permeable, and these in some instances might reflect institutional identities or organisations with regularised linkages.

As far as the application of systems concepts in decision-making analysis and theories of organisation are concerned, attention is often drawn to the following areas. Communication *patterns* may be seen as a central element, in as much as all organisations are alike in that they are held together by communication and with more than one organisation involved networks of communications may exist.[28] Where decision-making is specifically the focus, then the *content* of communications, and other forms of information, will also be of primary importance. With communication we must, therefore, be aware of both the patterns of communication between individuals, groups and organisations, and the explicit content of important 'messages'. We need also direct some attention to the extent that there are processes of learning and adaptation involved, through the mechanism of 'feedback'.[29] Both communications and feedback will have to be assessed in respect of the technical efficiency of these features and in respect of any political questions that may arise. It might be expected that communications between like-minded groups were important in arriving at concerted efforts in the wider policy debate. Similarly feedback could be expected to include technical as well as other responses of a more subjective form. Politics is about *relationships*, and for any individual group or organisation communications and feedback are important in political interactions. In this context, feedback can include both supportive ('positive') and unsupportive ('negative') information. But interesting questions arise from that. To what extent do organised groups actually search for information that is not supportive? Do they tend to display bias in the selective filtering of information from their environment in order to avoid disruptions to their policy choices?

Organisations, when faced with particular problems, usually take decisions in a fashion that is not as clear-cut as rationality might suggest, enabling bias to be displayed amongst options. None the less, decision-taking and implementation may involve activities of 'search' where alternatives are considered. Indeed, the alternatives may be suggested from both within or outside the organisation concerned.[30] Relatively simple rules may apply generally, where problem symptoms are identified and compared with the currently held alternatives. It has even been suggested that 'Organisations may engage in elaborate processes to justify a decision that has already been made.'[31] Such factors might create sources of potential disinformation, or veils through which the analyst must attempt to locate the source of decisions. In one of the more obvious examples concerning genetic manipulation, a charge was made against the US National Institutes of Health (NIH) that, in its assessment of potential courses of action, it never seriously considered a long-term ban on the research. In effect, the charge was that the NIH was working to a prior assumption that the research would continue, the question merely being how. Further, it appears that it took threats of legal action to persuade the NIH to fulfil its statutory obligation to produce a comprehensive study of the 'environmental impact' of putting its decision on guidelines into practice, an exercise which was required to address alternative courses of action. It only provided this study *after* it had actually issued the guidelines under which the research could proceed.[32] The difficulty lies in identifying where biases affected decisions, where limited search procedures operated and where suggestions of alternatives came from. This is especially so in a transnational analysis.

An example of the problem arises from the nature of the relevant science community. Within science communities, interactions and communication between groups and agencies are at high levels, but usually from a similar 'science' viewpoint.[33] Thus, alternatives suggested by certain like-minded organisations may not be 'real' alternatives, but suggestions around a preconceived set of ideas, which in fact reinforce those ideas. Such observations could apply to the behavioural characteristics of all types of organised groups, but they are particularly significant in activities that are politically controversial in the eyes of other groups. Access to decision processes is therefore often a key objective of groups with different perspectives and goals. At a basic level, this is often reflected in the division between those who hope to gain from a particular course of action and those who fear they will lose out, even though, as stated earlier, perceptions of outcomes and alternatives might be flawed all round.

Generally speaking, it could be said that organisations, in taking decisions or developing policy procedures, try to avoid the problems of uncertain information and knowledge. Graham Allison, for example, notes that in seeking to avoid uncertainty organisations follow two rules:

The first rule is: solve pressing problems rather than developing long-run strategies. The requirement that events in the distant future be anticipated is avoided by using decision rules that emphasise short-run feedback. The second rule is: negotiate with the environment. The requirement that future reactions of other parts of the environment be anticipated is avoided by imposing plans, standard operating procedures, industry traditions and uncertainty-absorbing contracts.[34]

Given the uncertain risks that were conjectured to surround genetic-manipulation techniques and the equal uncertainty about the most appropriate experimental guidelines to issue, it might be expected that efforts were made to frame policy questions in more familiar ways. However, we need to distinguish carefully between newly organised groups, which would lack any existing 'standard operating procedures', and more established groups. In point of fact, the avoidance of uncertainty, and the above features of searching for alternatives, go hand in hand with models of decision-making based on 'limited', or 'bounded', rationality and the seeking of satisfactory rather than optimal outcomes.[35]

Rather than searching for the optimising course of action relative to organisational goals, the organisation content with a satisfactory outcome will follow incremental patterns of choice, where the first found option that satisfies will be adopted. Limits to human and organisational perception in turn are likely to be reflected in the way that complex problems are broken down. Within organised groups, this implies the compartmentalisation of the problem amongst various subgroups. It may also be the case that compartmentalisation, or division of labour, occurs between different organisations, even internationally, if the initial complex problem is common to them. Thus any rationality that is evident in decision-making will be at best within the 'bounds' of the partitioning of the problem.

Where numerous organised groups are actively communicating, some attention should perhaps be directed towards key individuals, who play a role in more than one group, and who can have above average influence. W.M. Evans, who has developed a model of 'interorganisational relations', suggests that such 'boundary personnel' should be investigated in terms of their numbers, background and expertise, positions in the hierarchical structure, and normative orientation.[36] Various categories, or patterns of interaction, can be described in Evans' model centred on the prominence of a 'focal organisation'. A *chain network* most accurately describes the format of the example in Figure 1.5, whereas a *wheel network*, Evans suggests, locates the focal organisation at the centre of a number of other organisations, none of which are linked except through the focal organisation. Alternatively, an *all-channel network* implies more of a free-for-all, where all the organisations interact with each other. In examining the issues surrounding the emergence of the new biotechnology it is useful to be able to identify the most important domestic and international organisations, and the key individuals occupying positions on the boundaries of communication between them. As regards the more politicised issues, it is interesting to identify crucial centres where information is compiled and more importantly assessed, prior

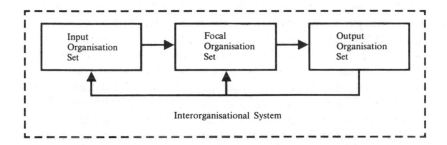

1.5 A model of an interorganisational system (From W.M. Evans, *Organization Theory: Structures, Systems and Environments*, John Wiley, London, New York, 1976, p.151)

to further dissemination. In effect, some organisations and individuals may have played roles of great significance, in establishing transnationally accepted norms, in the context of genetic manipulation.

In describing the international responses to the emergence of conjectured fears over genetic manipulation, the following chapters will give attention to the kind of factors suggested by systems thinking, as outlined above. However, critical as those features are, communications, feedback, organisational search, and the tendency of organisations to avoid uncertainty, need to be related to the more obviously political aspects. Perhaps the most obvious political question centres on who participates in the making of policy decisions in controversial issue areas. But participation need not imply the direct involvement of all individuals who display concern, but can mean the representation of their views in the decision process. Traditionally, participation in decision-making involving contentious issues has been related to the concept of 'power', and traditionally, in academic fields, disagreement over the precise nature of power has been endemic.[37] In the context of decisions involving areas of conflicting values, power and participation are related to the political achievement of the goals of actors. Participants wish to see their values as predominant in the outcomes, an observation pertinent to the genetic manipulation controversy.

To be broadly accepted, both decisions and the processes by which they are made need to be seen as legitimate by all who have a perceived interest. In the last resort this implies social accountability. This does not mean that every detail of a decision output should be acceptable to all concerned, but rather that any compromises or trade-offs of values are acceptable on the basis of general recognition of the validity of how they came about. Ideal as this might seem, there are other considerations deserving attention. First, it will be necessary to assess whether any actors applied power or influence in order to restrict the scope of decision-making to relatively 'safe' issues, in their view. Secondly, it may be expected that both organisational and political activity can be examined in order to determine just how issues entered the agenda of decision-making, and the overall politicised debate. These are important points to be borne in mind when appraising the genetic manipulation debate, not least because of the relatively technical nature of information pertaining to perceptions of both risk and benefit. Participation and legitimacy both involved the need to have technical and scientific information explained sufficiently well for non-scientist participants. Expertise in these areas could potentially endow power to those in its possession, or could raise distrust if observers of the decision-making activity questioned its legitimacy. As an example, at least one significant issue was minimised in most decision forums; namely the possible application of the new techniques in developing biological weapons.

A third insight suggests that: 'All forms of political organisations have a bias in favour of the exploitation of some kinds of conflict and the suppression of others because *organization is the mobilization of bias*.'[38] Although many of the organisations caught up in the genetic manipulation issue area were primarily concerned with aspects of scientific research, they found themselves entangled in a politicised debate. In so far as they directly made decisions, or supplied inputs into decision-systems, then organisational bias, representing the shared perceptions of the membership, was always a possibility. Taking these points a little further, a fourth focus might be on the likelihood of '*non*-decision-making' as a characteristic of the activity of some organisations.[39] That is, important decision options are avoided through mobilising bias, perhaps supported by power, defining issues as outside the scope of inquiry, blocking challenges to the prevailing consensus, and taking active measures to reinforce certain values. It will undoubtedly be difficult to prove any such activity but nonetheless indicators perhaps exist. However, one source of evidence for non-decisions could be the degree of grievance held by those who were disfavoured as a result of perceived non-decisions or who were excluded from the decision process.

Perhaps the most fundamental political requirement faced by the decision-makers in the case of genetic manipulation, as in most aspects of political life, was that of legitimacy. For authorities to be accepted, they must be legitimate, which implies that authority derives from those to which it is addressed. Legitimacy, therefore, involves reciprocal relationships, and in systems terms this suggests that loyalties would be directed towards the roles (representing values) played by participants. That is to say, participants a party to the various decision forums represent sets of values to which those not directly involved can at least focus loyalty. However, problems arise when participants play roles which are not legitimised in this fashion, which in turn may also lead to what Burton has termed 'role defence'.[40] Individuals or groups might try to remain in authoritative positions through, it is postulated here, activities akin to narrowing decision frameworks or through the mobilisation of organisational bias.

Because of the importance of information under conditions of uncertainty, the authoritative standing of the source of information cannot be ignored. It was after all the authoritative statement of a number of scientists expressing their concern over conjectured risks that opened the genetic manipulation debate. It is also clear that as the controversy unfolded, a number of non-scientists began to question the legitimacy of leaving decision-making to scientists, including those who initiated the concern. Such shifting of the perception of legitimate authority requires explanation. It also necessitates a degree of comparative analysis between the different national decision-systems as well as an understanding of how they are interlinked.

The case of genetic manipulation will undoubtedly be of interest to future historians looking back on a remarkable event in the history of science. One historian, Charles Weiner, foreseeing the importance of the story, which began to reach its zenith in the mid-1970s, established an 'oral history' archive at the Massachusetts Institute of Technology. Weiner and his colleagues embarked upon an extensive series of interviews and document-collecting covering many of the important individuals and organisations central to that story. More than 120 interviews were recorded, many transcribed, and 1700 letters, 1500 documents and 11,000 articles were collected and catalogued by 1978.[41] But as another historian has observed:

It is the historian who has decided for his own reasons that Caesar's crossing of that petty stream, the Rubicon, is a fact of history, whereas the crossing of the Rubicon by millions of other people before or since interests nobody at all.[42]

The wealth of material available in documenting the events and science surrounding the emergence of the new biotechnology is enormous, and it is with the Rubicon lesson in mind that I have made explicit the concepts selected with a view of providing a framework of analysis. The framework may or may not have its problems in the eyes of other readers, but I hope the insights derived from its application to the story set out in the following chapters will be fruitful nevertheless.

2 Dimensions of Risk and Benefit:
The Origins of International Concern about Genetic Manipulation

At the height of anxiety about the new techniques of genetic manipulation there was a dearth of quantifiable information to suggest hazards if the techniques were used. Consequently, the approach adopted here will be to demonstrate that although concern rested on conjecture, it came from authoritative sources. Thus there is a premise that knowledge, in this case about hazards, is not solely derived from empirical evidence, but is drawn from theoretical explanation founded upon both empiricism and logical deductions.[1] It was deductive logic which underlay the initial caution as potential hazards became conjectured. Hypotheses about *uncertain* risks were the result, which eventually faced tests, either designed specifically or brought to attention through the passage of time and the experience and knowledge gained from the use of the very techniques themselves. In order to demonstrate the importance and legitimacy of the early disquiet, a descriptive and historical summary is presented, indicating the fashion by which attention became drawn to potential hazards. In particular, the events behind the publishing of two influential letters and the holding of a now famous international conference in California deserve special attention. Together these were responsible for initiating responses in many states and international organisations.

THE BERG EXPERIMENT

An experiment which Professor Paul Berg of Stanford

University proposed to do in 1971 has been commonly suggested as the first step in a chain of activity leading to the international discussion of potential hazards and the development of experimental guidelines. He had intended to implant the DNA from Simian Virus 40 (SV40) into the bacterium *E. coli*, a normal inhabitant of the human digestive tract. Attention became drawn to this when a graduate student from Berg's laboratory, Janet Mertz, attended a course at Cold Spring Harbor Laboratory, New York. There, Robert Pollack gave a lecture which included discussion on matters such as safety and ethics in work with mammalian cell cultures. In particular, Pollack was concerned with experiments to study viruses, cancer and the like. Not least, he referred to some of the less well-known hazards involved in indiscriminately mixing cells of different species under artificial conditions.[2]

Mertz described Berg's proposed experiment. SV40 was a virus known to cause tumours if injected into newborn hamsters, although was thought to be safe in humans.[3] In general, viruses associated with causing tumours were of particular interest to those who hoped to find a viral component of human cancer. Pollack began to wonder about the risks in introducing SV40 into *E. coli*, which could then perhaps produce the virus in large quantities. Although apparently safe in humans, SV40 might find itself transmitted more readily in *E. coli*, and if introduced into humans, the *E. coli*, reproducing and carrying the virus, might have long-term effects. Berg has acknowledged in interview that he took some convincing from Pollack and Mertz that his experiment would be risky.[4] Consulting many other scientists, Berg found a number were very critical, including Maxine Singer and David Baltimore, who were later to become important in publishing concerns. Still uneasy, Berg sought advice in the summer of 1972 in a lecture at the European Molecular Biology Organisation (EMBO). After a lengthy session, it seemed that no one was sufficiently confident or knowledgeable to give him his answers.[5] In subsequently abandoning the experiment, although partly because Berg saw inherent difficulties in it (the virus might not be expressed in the *E. coli*), Pollack felt, much to his relief, that Berg's action would deter others from similar experiments.

From this modest beginning, discussion of potential hazards in biological research gained momentum around the world. A series of conferences and meetings gave vent to the issues. Andrew Lewis, of the US National Institute of Allergy and Infectious Disease (NIAID) presenting a paper on SV40 hybrids (at a symposium at Cold Spring Harbor, immediately following the workshop in which Pollack and Mertz had been involved) again raised the question of risk.[6] A yearly NATO meeting convening in 1972 in Sicily, after a lecture by Berg, gave over an evening to discussing the political and social consequences of genetic engineering.[7] Later in the year, an EMBO workshop held in Basle on DNA restriction and modification gave another evening over to issues surrounding the creation of genetic hybrids. November 1972 saw the establishment by the US National Institutes of Health (NIH) of a Biohazards Committee, and Andrew Lewis initiated action on the part of NIAID to control the supply and use of possibly hazardous hybrid viruses. Lewis was responding to hazards he saw in his own work, but it became NIAID policy that those wanting virus samples would have to sign a memorandum of understanding in which they agreed to take certain safety measures. If they passed the virus to anyone else, the recipient should be required to promise likewise.[8] In general, these moves were paralleling the development of the methods of cutting and joining DNA through the use of restriction enzymes.[9]

By now Berg had become convinced that there was a question of biohazards to face up to, and in conjunction with others he was to take events further. A series of special conferences was proposed, the first to be a fact-gathering exercise on tumour viruses and at least one other to address types of experiment being proposed and the associated hazards. Berg asked two colleagues, A. Hellman and M.N. Oxman, to assist him in organising the first conference specifically on biohazards. Sponsored by the US National Science Foundation, the National Cancer Institute and the American Cancer Society, it was held between 22 and 24 January 1973, at the Asilomar Conference Center, Pacific Grove, California. This was to be the same venue as a later and much more important international conference (for convenience they can be termed Asilomar I and Asilomar II).[10] Essentially the discussion at Asilomar I was on biohazards in general rather than on recombinant DNA.[11] Later more specific concerns would be addressed. One question, however, would remain important in genetic manipulation. It was noted that the background training of those dealing with virology was all too often in biochemistry, and a tendency existed for viruses to be seen as yet another chemical reagent.[12] Other issues examined included: laboratory infection from animals; evaluation of experimental results; tumour viruses; hazards; modifications to viruses; common sense in the laboratory; and the control of hazards in cancer research. It was noticeable that no hard evidence was presented to prove conclusively specific hazards, yet an awareness of overall hazard was fostered. Equally of note was that ethical and moral questions were not raised, as they had been at the NATO meeting. Nevertheless, the potential hazards to health were taken seriously as James Watson, co-discoverer of the structure of DNA, observed at the meeting:

Of course everyone working with a given virus hopes very much that his particular virus is indeed safe. But I think we must now help create the situation where the real reason for a decision [for biohazard prevention] is the awfulness of the alternative possibility–which I suspect is how the AEC calculates the low probability of a catastrophic accident to a nuclear power plant.[13]

In sum, the development of discussion on biohazards, following Pollack's worries over the experiment which Berg proposed, provided a background to which specific concern over recombinant DNA techniques was later voiced. The next significant development was the publication of two letters.

THE SINGER–SÖLL AND BERG LETTERS

By the spring of 1973, Stanley Cohen, Herbert Boyer and some colleagues had succeeded in inserting DNA from one plasmid into another using the restriction enzyme *Eco* RI. Their main contribution was finding a plasmid which maintained its replication functions even with the DNA inserts.[14] Their work was published in November 1973,[15] and it attracted much interest. The technique spread quickly as Cohen made the plasmid which had been so useful available to other researchers. Recognising the question of hazards, he requested that they made assurances that tumour viruses would not be inserted into it, or any other DNA which might make the *E. coli* carrier of the plasmid more resistant to antibiotics. He requested that they should not pass the plasmid to anyone else without similar assurances. However, this self-regulated control broke down as other useful plasmids were discovered elsewhere.

In the same year a conference was held, this time explicitly addressing, in a special session, the newly developed techniques. More importantly, the participants voted for a letter to be sent to the National Academy of Sciences (NAS) and the journal *Science*. The conference was the annual Gordon Conference on Nucleic Acids, and the vote followed a special session called for by participants after they had heard Boyer talk about the new techniques. Maxine Singer opened the session from the chair, noting the potential for the techniques and that moral and ethical issues were involved. The scientists were then told of their responsibility to co-workers, laboratory personnel and the safety of the public.[16] Following the discussion, a vote was taken in which 78 of some 90 participants voted in favour of the two chairpersons sending a letter to the NAS. A significantly narrower vote of 48 to 42 decided in favour of giving the letter wider circulation through publication.[17] As it has come to be known, the Singer–Söll letter addressed a matter of 'great concern'–the joining of DNA from diverse sources. Briefly describing the new techniques,[18] they noted that hazards, although not established, could be potentially involved. Referring to the fact that the letter was voted upon, Singer and Dieter Söll said that the conference was calling upon the NAS to establish a study committee to recommend 'specific action or guidelines, should that seem appropriate'. The response was very prompt, and in October 1973, Singer was invited to a meeting of the Executive Committee of the Division of Life Sciences at the NAS,[19] where it was decided to proceed with NAS involvement. On

Singer's suggestion, Berg was asked to advise them on the basis of his earlier actions. A study committee resulted. After consulting Watson and J. Edsall (an 'elder statesman of biochemistry', as he described him), Berg decided to call a meeting of those in the field.[20] Eight scientists subsequently attended a meeting at the Massachusetts Institute of Technology (MIT) in April 1974, although Singer, unavoidably detained at the last minute, played no part in drafting the more important letter which eventually resulted from this group.[21] Berg revived his earlier idea of a large conference,[22] although the meeting was not sure whether or not to go ahead with this. Nevertheless, Berg booked the Asilomar centre in case it was needed for February 1975, a time when it was free. To Berg some experiment proposals he was hearing of were 'very worrying'.

Agreement from Berg's colleagues was forthcoming and planning for the meeting began. In the meantime, Norton Zinder made a suggestion that if they 'had any guts at all' they would request researchers to halt their experiments. Watson has attributed part of this concern to the 'mysticism' surrounding the way tumours could result from certain viruses and the discovery of human viruses similar to SV40.[23] It was decided to draft a letter, although a number of arguments were put regarding how forceful their request to halt any work should be. Of importance is the order to the two decisions made that day. The decision to hold a conference was agreed upon before the idea of a letter, which was seen to be a stop-gap measure.[24] Both must be taken together. The conference when held was not a consequence of the letter as is often implied,[25] although the meeting undoubtedly received stimulus from it.

Often criticised in retrospect by some members of the scientific community as a hasty document, the 'Berg letter', as it came to be known, was on the contrary not so. It went through a series of drafts over a period of two and a half months, and it was widely discussed in the US and abroad. Despite its common label, the first draft was produced two days after the MIT meeting by Richard Roblin, and was sent to Watson and Baltimore for comment, before Roblin redrafted it. On 20 May, Roblin produced what was hoped to be the final

version, which was a composite of the second draft and
versions produced by Nathans and Berg. It was sent to all who
had been at MIT plus the NAS (who had financed their
meeting), Stanley Cohen, Herbert Boyer, David Hogness and
Ronald Davis. Philip Handler, the NAS president, was,
however, unhappy as it did not make clear the NAS role. Thus
the president of this prestigious institution rewrote it!
Nevertheless, Berg *et al.* disagreed with the latest version,
arguing that it made their requests appear to be an edict from
the NAS.[26] Berg himself had felt that to have impact the letter
should appear as a personal appeal from the scientists. After
an abortive attempt by Roblin to include Handler's
suggestions, they met the latter for discussion. It seemed that
Handler had the impression that Berg's group was an NAS
committee, a view they disputed, arguing that none of them
saw themselves as a committee, and in any case had expressly
been told not to consider themselves as such. Finally, Berg and
Handler produced the version actually published, which
included a paragraph saying that the NAS had invited the
group to meet. During the months of drafting, opportunity for
wider discussion of the letter had existed, including a meeting
of EMBO at Ghent, Belgium, convened in May 1974 to discuss
restriction enzymes and nucleotide sequencing. At this,
Nathans and Zinder presented the idea of the statement, to
which there was general agreement. In the United States,
Baltimore read the tentative draft to a symposium on tumour
viruses at Cold Spring Harbor, in June. Thus, as Weiner
observed, 'the relevant scientific community knew what was
up.'[27] Indeed, following its presentation at New York, twelve
European scientists wrote to John Kendrew, the Secretary-
General of EMBO, requesting urgent consideration of the
matter, and in particular the provision of a special risk
laboratory.[28] Therefore, by the time of its publication in July
1974, the letter had been well considered internationally.[29]

In brief, the Berg letter requested the deferment of two types
of experiment until the hazards could be evaluated: first, the
construction of new self-replicating bacterial plasmids that
might introduce antibiotic resistances to plasmids which do
not have them, or introduce novel combinations of resistances;
secondly, the linkage of DNA from oncogenic (tumour-

causing) or other animal viruses to self-replicating DNA elements such as plasmids or further viral DNAs. It was also advised that great care be taken in linking any animal DNA to plasmid or bacteriophage DNA, because of the uncertainty involved in creating any new recombinant DNA molecules whose properties would be difficult to predict. To the director of the NIH they requested the establishment of an advisory committee charged with overseeing an experimental programme to estimate hazards; developing procedures to minimise the spread of such molecules in human and animal populations; and devising guidelines for investigators to use. Finally, they requested that an *international*[30] meeting of involved scientists be held early in the new year (the venue having already been booked). Nothing like the publication of such a set of requests had occurred in the history of biological science.[31] Other techniques or scientific developments may have displayed risks sufficient to suggest caution, but none in recent years have had debate, subsequently politicised, begun in such dramatic form. All of this had occurred in the case of recombinant DNA, before any hazards were even proved. Because risk could be conjectured authoritative scientists had requested important measures.

The period following the Berg letter has become known as the 'moratorium', a term Berg believed arose from the press. In conjunction with their letter, Berg's MIT group had decided that it would hold a press conference on 18 July to stress their requests. Between 50 and 60 reporters attended. Although Berg himself was critical of the strict interpretation of the letter as a call for a moratorium,[32] this does not in any way detract from the fact that, as Watson reflected in 1979, the letter was indeed very strong.[33] Such a letter could not help but be newsworthy, and even if the authors did not anticipate the response that over the years would develop, this in no way reduces their authority. With responses to the letter so widespread internationally, and the NIH responding to the requests made of it, it would not be surprising if the press had found their own story. Press sensationalism, however, is a charge that had more bearing in the United States, while the responses to the letter were more international. For example, in the United Kingdom, also a very important respondent in an international sense, the popular press never took much active interest in the ensuing issues. The politics of congressional lobbying in the United States may explain partly press interest and the higher profile of public debate on that side of the Atlantic.

On the basis of the actions so far described, a worldwide voluntary deferral of certain experiments, no matter how unlikely this might appear to sceptics, did in fact occur. In retrospect, the NIH has argued that evidence for this success comes from both informal communications in the field, and from the inspection of publications in scientific journals.[34] Some scientists even suggested the deferral of those experiments involving animal DNA over which Berg's group had advocated caution.[35] A Japanese scientist, Kohji Hasunama, for example, who wrote to Berg fully agreeing with the letter was one. Roy Curtiss, an American, took it upon himself to raise the standard of caution even further. Drafting a sixteen-page single-spaced memorandum, he sent it to about a thousand scientists around the world. Again agreeing to the principles of the Berg letter, he made technical suggestions to widen the categories to four types of experiment to be deferred.[36] Indeed, in the United Kingdom, the Medical Research Council (MRC) did put into operation a complete ban on the third type of experiment, in addition to those recommended in the letter, for all the work which they sponsored.

In general, the point to note is that caution became endemic amongst those scientists who might plan to use the new recombinant DNA techniques, a credit to the standing of those who signed the Berg letter. But deferring the work was only one of the requests that had been made. It was always intended that the work should continue as soon as possible, under appropriate precautions. Risk assessment was asked for, as was an NIH committee to oversee progress from that point on, at least as far as the US was concerned. Yet, even before publication of the letter, the then Director of the NIH, R.S. Stone, indicated to Handler at the NAS that a committee would be formed, and that the NIH would support the international meeting.[37] In the US, the NIH is the main federal government agency for the conducting and funding of biomedical research, of which microbiology is a part. On 7 October 1974, the promised committee was established, with the ungainly title of the Recombinant DNA Molecule Program Advisory Committee. In later years it became more commonly known as the Recombinant DNA Advisory Committee (RAC) and, for simplicity, this latter address will be used here.[38] The committee was to advise the Secretary, Health Education and Welfare (HEW), the Assistant Secretary for Health, and the Director, NIH, on three defined functions: 'the evaluation of potential biological and ecological hazards of DNA recombinant of various types'; minimising the spread of such molecules; and devising guidelines to be followed by investigators 'working with potentially hazardous recombinants'.[39] Its Charter described the RAC as a 'technical committee', which reinforced the early dominance of the health hazard issue, within institutional responses, rather than wider issues. To support this, the membership of twelve was to be drawn from the fields of molecular biology, virology, genetics and microbiology. It would be some years before non-scientist representation would appear on this committee. The next chapter will outline the role of the RAC and criticisms against it in more detail. However, it is worth stating at this point that it was to become very influential both within the US and in giving advice, or as a precedent to follow, abroad.

In addition to the MRC-imposed deferments on experiments, the UK was also quick to respond to the Berg letter by establishing a working party under the auspices of the Advisory Board for the Research Councils (ABRC). Again this was announced a week after the letter appeared in *Nature*, and on the day of its US publication. From the Ashby working party, the UK response would begin proper. Thus, on the day of US publication of the fundamentally important Berg letter, key chains of activity began on both sides of the Atlantic and, as with the US response, the procedures adopted in the UK would act as a model to influence other states. It is abundantly clear, however, that Paul Berg and his co-signatories had no conception of the sheer extent of the eventual consequencs of their action. From the origins of concern within a relatively small scientific community, the subsequent debate over genetic manipulation would go through various stages domestically and internationally. More scientists would become involved, public interest would grow, controls would develop, legislation would loom and increasingly the original small band of scientists with their honourable sense of duty would begin to wish they had acted differently. In short, their actions had both scientific and political consequences. As Stanley Cohen was to say in 1979:

in retrospect, it seems to me that while the letter was *perceived* as responsible, it was not really responsible at all. The most incriminating thing that any of us could have said at the time about recombinant DNA research was not that there was any indication of hazard, not that there was even any valid scientific basis for anticipating a hazard, but simply that we could not say with certainty that there was not a hazard. The same thing could have been said about virtually any other kind of experimental endeavour.[40]

The extent of consultation and thought underlying the Berg letter has already been indicated. Cohen's comment, typifying similar views held by other scientists by 1979, needs response on the basis of its logic. First, Cohen underestimated the extent of deductive reasoning that was applied to the empirical knowledge of the day. The Berg letter was a *reasoned* appeal. Secondly, and perhaps more subtly, he was in effect saying that the hypothesis that there might be risk, at the time, could not be falsified. There are philosophical arguments to suggest that the latter is preferable as a statement to a continued collection of inductive evidence supporting an hypothesis. Even by 1979 the hypothesis of hazard was not resolved, although there was powerful motivation indeed to try to falsify it. Finally, the Berg letter itself was actually calling for research to obtain the necessary information to determine the extent of risk. It is not the intent here to enter into discourse on the relative merits of differing methods of scientific investigation. The point is that comments like Cohen's in general were much influenced by the failure of risk to manifest itself, and the acquisition over the years of new knowledge.[41] Cohen's observation is too simple. Something cannot be irresponsible simply because in retrospect the conjectured reasons for caution have proved less evident. Any action must be related to perception of the time. As Cohen says, at the time, the action was perceived as responsible. Perceptions can change, but later additions to knowledge, which modify perceptions, are not sufficient reason to challenge the responsibility of actions under earlier perceptions. Cohen must have perceived his signing of the letter at the time as responsible.

Given the authority attributed to the letter at the time, the next major event was responsible for extending the number of scientists involved, and the extent of internationalisation of the issues. The international conference proposed by the Berg group was in this sense a landmark. With internationalisation and even greater publicity, the politics proper began. Despite other options, all the proposals in the letter were followed.

THE ASILOMAR CONFERENCE, FEBRUARY 1975

In late February 1975, 150 participants attended the international Asilomar II conference. Organisation, however, had begun on 10 September 1974 at a meeting held at MIT.[42] Some initial decisions on participation were of note: two Europeans were to be invited to join the organising committee;[43] individuals were to be invited on their own merits rather than as representatives of any body, as it was hoped to avoid the meeting appearing political; the meeting was to concentrate on the science and technology in relation to health, and the participants would reflect that in being nearly all active scientists, from a variety of backgrounds;[44] experts on infectious disease, immunology and gastroenterology were to be present; to provide a wider input, a number of lawyers were to be invited, including Maxine Singer's husband; the press were to be invited, but with the novel provisos that they registered as attendees and that they agreed only to publish after completion of the conference agenda, to avoid hasty reporting.[45] Thus, in terms of participation, the emphasis was on scientists, with only four lawyers and the press to complement this.

Some technical decisions were also made on the same day, which provided the structure for the conference agenda. Three working groups were proposed, which would meet prior to the conference and finally submit a report to the attendees. A *Bacterial and Plasmid* working group would look at the biology of these in terms of the introduction and transmission of drug resistances, general epidemiology and similar questions. A *Viruses and Viral DNA* group was to examine animal viral DNAs, virus fragmentation, SV40 hybrids, relationships with tumours and immunology aspects. The third group was to examine *Eukaryotic DNA* and consider animal gene transfer and amplification, summarise pertinent work, assess risks and consider the advantages of actually doing hazardous experiments. It was hoped that with these groups, most concerns of the day would be covered.[46] That is, most technical issues of risk.

It is worth at this point saying something of the activity that occurred between September and February, when the conference met, which furthered interest in the biohazards that might be involved in genetic manipulation. In the same month as the meeting of organisers, an international organisation of note expressed positive interest by holding a meeting in Tokyo to discuss biohazards. The International Association of Microbiological Societies (IAMS) established an *ad hoc* committee which began with groups in different states.[47] Contact was established with members of Berg's group and the Ashby working party. It was important as the first international organisation to establish procedures to monitor and assess issues surrounding recombinant DNA, and reflects growing international awareness. Less technical, and more ethical, issues were, however, addressed in one particular meeting, despite the trend to narrow the focus towards the science in most discussion forums, including Asilomar II. In October 1974, in Davos, Switzerland, an international meeting was called explicitly to look at wider concerns associated with recombinant DNA work.[48] Although the meeting lacked coherent direction and the structure of discussion degenerated somewhat,[49] at least it established wider interests. Directly as a result of the Berg letter, the Davos meeting never really found its target in terms of establishing a conference view on the issues. Paul Berg, in attendance, made it clear that he was not interested in ethical issues, but only public health, while other participants either wanted to curb generally unbridled scientific research or wanted to remove all controls. Whatever its failings, it made some impact, if not directly, then through its published proceedings. Its chairman, H. Wheeler, writing after the event, criticised both the Berg conference and the Ashby working party for having insufficient non-scientist representation. He observed that the problems were social as well as scientific, and that a lost opportunity had presented itself at Davos where both biologists and social scientists had met. In some ways it was unfortunate that the scientists, about to embark on self-regulation at Asilomar, had not widened their horizons, as they might have realised the full extent of what they were unleashing.[50] Issues discussed at Davos would be raised time and time again as participation in the debate widened to include non-scientists.

A final development of note which was timed deliberately to occur prior to the forthcoming Asilomar II meeting was the publication of the Ashby report in the UK. Published a month beforehand, it was planned to be available to represent something of a British input to the forthcoming discussions.[51] Ashby hoped that the report would act as a consultative document in the UK, and it was acknowledged as a first attempt to respond to the issues raised by Berg's group. Of note is that the report recommended that means be devised to enable the work to continue as soon as possible, in that great benefits could result.

Eventually, the Asilomar II meeting convened between 23 and 27 February in the very pleasant setting of Pacific Grove, California.[52] Four main issues can be identified as having dominated the proceedings.[53] First, the meeting had to come to terms with what it was trying to achieve. Secondly, a major theme was the desirability and content of guidelines for use by investigators. Thirdly, debate arose over whether or not some experiments should be deemed too dangerous to undertake under any circumstances. Fourthly, there was a proposal that enfeeblement of biological hosts might overcome many problems of containing risk. These can be taken in turn.

Establishing the aims of the conference was partly determined by the prior decisions calculated to restrict discussion essentially to technical questions. Indeed, it has been said that the conference perpetuated the assumption that the analysis of risk was 'narrowly technical in nature'. Broad public and scientific input was not considered.[54] It was, however, hoped that a 'consensus' of those at the meeting would be forthcoming, assisted by the choice of invitees. David Baltimore, opening the conference, stressed the importance of avoiding splits which would infer that they had failed in their duty. Uncertainty is nevertheless revealed in his comment that 'the procedures by which the consensus will be determined will be largely determined by the extent of the consensus'.[55] It was clear that it was hoped that any consensus would include some form of self-imposed guidelines under which the seven-month moratorium could end. This in itself would be something of a landmark in the history of science. Particularly influential at the meeting, Sydney Brenner, the British member of the organising committee, pressed for guidelines so tight that no one could accuse them of being self-serving. A good guideline would be one in future revised downwards.[56]

Two voices were notable, though, in their arguments against any regulation, and both were Nobel laureates. Joshua Lederberg and, perhaps surprisingly, James Watson, argued respectively that delays to research would delay benefits accruing and that academic freedom was at stake. Watson went so far as to recommend the use of common sense and 'live with the fact that someone may sue you for $1 million if you are careless'.[57] Their fears were shared by others in that their drafting of regulations could subsequently lead to legislation.

Guidelines were produced out of Asilomar II, although discussion of their content was long, and revolved around the reports submitted by the three working groups. From the Bacterial and Plasmid group came the suggestion of six classifications of experiment to which physical containment recommendations were attached. It was during this debate that Brenner made his defence of guidelines and Watson and Lederberg raised their reservations. The Viruses and Viral DNA group produced the controversial suggestion that existing National Cancer Institute guidelines for handling oncogenic viruses would be quite sufficient. Andrew Lewis of NIAID fuelled the debate by recounting his experiences with trying to get people to impose self-regulation when he supplied SV40. Some passed on the virus without obtaining guarantees on its use. Such criticism of self-regulation was not popular, given the evolving aims of Asilomar II. Yet more controversy followed the report of the Eukaryotic group. Suggesting that research with eukaryotic DNA might be the most fruitful of all, they proceeded to outline their fears of the 'shotgun experiment'.

Viewed as potentially quite frightening, the 'shotgun experiment' involved using a restriction enzyme to fragment a strand of DNA, the fragments then to be cloned in a host–vector system. Each fragment might contain two or three genes, each coding for the production of particular enzymes or proteins. This sort of experiment it was thought would be popular as the isolation and categorisation of mammalian genes was likely to be of high priority in future microbiology. Uncertainty about the risks arose from the observation that certain segments of mammalian DNA appeared to be very similar to the structure of known tumour viruses. It was postulated that these segments were perhaps kept in check by their location relative to other genes in the strand. Isolated and placed in a bacterium they might be expressed. A further possibility was thought to be the introduction of an unknown foreign gene that happened by chance to code for toxin production, or fundamentally altered the bacterium itself.[58] The Eukaryotic group had put the 'shotgun experiment' on top of a hazard list within some guidelines, which led to wide discussion, but with problems as Wade, a journalist, noted, 'the central dilemma the experimenters faced was that, despite the various attempts to rank the experiments in order of risk, no one had any real idea of what the risk might be or how to assess it.'[59]

A question also arose at Asilomar II regarding whether or not certain restriction experiments should be totally avoided. After much discussion, it was decided that there were some, although not defined at that time.[60] The last major issue was related to all assessments of hazard in that it concerned the use of *E. coli*, which had been at the heart of perceptions of risk for some time. At Asilomar II, the British scientist E.S. Anderson suggested that the laboratory strain *E. coli* did not in fact survive long in the human gut. A productive session followed, where it was proposed that strains of *E. coli* could be bred such that they could not survive at all outside their culture medium. This would reduce reliance on physical safeguards which could always fail. It began to appear that guidelines involving biological as well as physical containment would be more attractive.[61]

Each of these were important technical issues, but two final sessions were important to the overall influence of the meeting. On the evening of the third day the lawyers spoke, with forceful impact. A number of issues were raised regarding risk assessment and participation. Singer and Capron attacked the idea of academic freedom if it could lead to harm to others, and they challenged the competence of the scientists to assign overall risk. Capron argued that the public had a right to act through the legislature and even to make 'erroneous decisions'.[62] However, the talk which had the scientists most worried was delivered by R. Dworkin. He raised the issue of legal liability, suggesting, for example, that the US Occupational Safety and Health Act might be applied to protect laboratory workers and the financial liability of institutions. By the end of the session the scientists were being advised to examine the possibilities of extended liability insurance!

Some scientists lost sleep that night for quite different reasons! The organising committee in search of a positive result from the meeting worked through the last night drafting a conference document. Synthesising earlier suggestions, they reduced the six categories of the plasmid group to four–minimal risk, low risk, moderate risk and high risk. Playing on their minds in addition to this was that the conference all along lacked means of registering opinions of the group at large. Voting was not in effect organised, and Berg in particular was reluctant to call a vote on their draft document. Berg hoped that a consensus would simply appear and be obvious. On the last day, however, a principle which they had drafted was put to the vote. It read:

The new techniques combining genetic information from very different organisms place us in an area of biology with many unknowns. It is this ignorance that has compelled us to conclude that it would be wise to exercise the utmost caution. Nevertheless the work should proceed with appropriate safeguards.[63]

When taken, the vote unanimously endorsed this principle as a first paragraph in a statement. During the morning, modifications and suggestions were made and voted upon, and discussion progressed to consider the body of the conference statement. Despite vocal objections during the past few days, it appeared that the silent majority approved of the actions of the organising committee. The committee began to think that the whole statement would be acceptable. A revealing note was passed from Maxine Singer to Berg. It read:

Paul: If you sense, as Dick and I do, that all the votes will go in favor of the statement, or, indeed a somewhat stronger statement, then there may be a lot to be gained by taking a vote.
We are *already* over the main hurdle, since we have overwhelming votes in favor of the *principle*.[64]

When moved, the vote was overwhelmingly in favour of the draft prepared that morning, although by then it had many qualifying statements along the lines of how 'provisional' it was, and that it was a 'first assessment'. It also suggested that investigators had a responsibility to increase containment if they felt that their experiment required it.[65]

A number of drafts were made after the meeting before publication of the final form in June 1975, in the US and the UK. An erratic and narrowly focused conference in the end produced an output of considerable influence, not least in enabling some research work to begin again. The statement produced guidelines relating hazard to type of biological 'material' used: prokaryotes, plasmids and bacteriophages at the low end of the scale, through animal viruses in the middle, to the use of eukaryotic DNA at the more hazardous end. Such self-regulation was unique on this scale. If the central argument of this book rests on the authoritative announcement of conjectured hazard, then the Asilomar statement, resulting from a conference of some 150 interested scientists, simply completes a *repeated* announcement of perceived potential hazards and uncertainty. In conjunction with the Singer–Söll and Berg letters, it provides conclusively a base upon which to examine the organisational responses to the expressions of concern made very early indeed in the development of this important technology. Creditable caution was the norm.

Perhaps justifying the strictures on when the press could report, their articles, produced on completion of the meeting's business, were on the whole sober and cogent descriptions and analyses. Weiner, commenting on their behaviour, observed:

At Asilomar, the press representatives were not allowed to write about the meeting until it was over. Instead, the scientists were the ones rushing to the telephones to tell their colleagues back at their own laboratories what was going on and to suggest ways to get involved in the new techniques.[66]

As the debate developed and international response grew, participation would widen to include an increasing proportion of non-scientists, subsequently challenging the scientists' control of the events. Their responsibility displayed so openly led inexorably to perceptions in some quarters that because scientists had acted thus, then there *must* be an underlying hazard. One group, the Boston Area Science for the People, had actually sent an open letter to the Asilomar meeting. They suggested that scientists alone could not make the decisions on the future direction of the research and that those at risk through working in the scientists' laboratories, plus the wider public, should have a say. They expressed worry at the research going ahead before all the requests of the Berg letter were met, in particular (and they had a good point here), the risk assessment called for. Also of interest, they requested wider participation in the RAC, which by then had been announced and met for the first time the day after the Asilomar conference.[67] Such issues will be returned to in later chapters.

GENETIC MANIPULATION AND BIOLOGICAL WEAPONS

Scientists themselves, as with the origins of overall concern, began the speculation about the utility of recombinant DNA techniques in assisting the production of biological weapons. Between 28 August and 2 September 1974, Pugwash held its annual conference in Baden, Austria. Under 'other business', and as a result of the Berg letter, Martin Kaplan, also involved in the World Health Organisation, presented a short paper informing those attending of the developments which had occurred in genetic manipulation.[68] In discussion following the paper, he and Ole Maaløe, from Denmark, presented the hypothesis that if a gene for a toxin could be isolated, it might be possible to use the *E. coli* host–vector system to produce large amounts of the toxin (assuming its successful expression) more efficiently than with fermentation methods.

Some 40–50 scientists and social scientists were present and their collective feeling was for the moratorium to be endorsed, for legitimate public interest to be represented and for their concerns about the use of genetic manipulation for biological weapons to be made known. In particular, they wanted these views brought to the attention of the Pugwash Continuing Committee, the NAS in the United States and the scientific community in general. The latter would involve the use of scientific journals and 'the usual channels used by Pugwash'.[69] It is important to note that Pugwash, and in particular Kaplan and Maaløe, had long been interested in the question of biological warfare. At the 1976 Pugwash annual conference the issues were again raised with a similar conclusion, perhaps if anything more strongly worded.[70]

If the Berg letter was important in raising general concerns about the hazards of recombinant DNA research, we can only speculate as to what its impact would have been if the following paragraph, included in two of the drafts, had been published:

Finally, since it is evident that these new technological capabilities could potentially be used to create new sophisticated weapons of biological warfare, we urge citizens, scientists and government officials to take appropriate steps to prevent such applications.[71]

Somewhat sceptical about biological warfare having worked at Fort Detrick, Watson later commented on the omission of this paragraph: 'this phrase was removed at a later draft because it would raise an issue which would involve unavailable classified data. Moreover, no one believed that if the CIA or the army wanted to do such work that our moratorium would stop them.'[72] The availability of classified data is an irrelevant excuse, as concern in 1974 was with conjectured uses for which no data could exist. A more understandable explanation for the omission of the statement was the Berg group's wish to avoid sensationalism.

Some concern was therefore evident regarding the use of the new techniques in biological weapons production. That it was played down by the scientists who raised the whole question of conjectured hazard is not surprising, as it would almost certainly have precipitated the involvement of many interest groups. It might also have brought uncomfortable comparisons with that earlier technology–nuclear energy.

SUMMARY

This chapter has traced the historical origin and development of concerns over the possible hazards associated with the microbiological techniques outlined in Chapter 1. It is clear that the Berg letter and the Asilomar II conference were both designed to have international prominence. Accordingly the actions which were taken in an authoritative fashion were not hasty and took account of discussions at various professional meetings, some of which were international. If Asilomar II marked the high point, then two and a half years of increasingly international discussion had led to that point.

However, there must be applied a sense of time in examining these early beginnings of concern. The events outlined above took place in a period of some distrust of science and the impacts of technology on the environment, particularly concerning the nuclear industry. In the words of Pringle and Spigelman: 'The nuclear community was a group under siege; yet, like Paracelsus, the zealots were if anything, more confident than ever.'[73] An increasingly organised anti-nuclear movement embraced the nuclear industry in heated debate at a time when much attention was being directed to environmental issues and the erosion of non-renewable resources. If man had apparently ransacked nature in the past, some people were to see genetic manipulation as the start of his ability to control the essence of life itself with untold consequences for nature. As controversy unfolded from the Berg group's expressions of concern, the pressure-group activists would find a new issue to focus upon. Many scientists themselves, particularly the younger ones, were not untouched by these viewpoints.

The responses of institutions in the US and the UK will now be treated in some detail, followed by a summary of the wider international impact, including other countries and emphasising the importance of certain international organisations.

3 Response to Concern: The US Model

Events in the United States must be considered to be of particular importance, not least because of the extent to which guidelines produced by the National Institutes of Health (NIH) were to influence those of other states. A report published in 1980, having surveyed membership of the International Council of Scientific Unions (ICSU), records 28 states with guidelines in operation, 15 of which had modified those of the US.[1] Because of the extent of adaptation of their guidelines, it is fruitful to examine in some detail the processes by which the US guidelines were developed and how they were operationalised. For this purpose an historical review of the decision-systems will be followed by a summary of the system involved in their implementation. Events in the US were by no means closed to the influence of foreign actors, but for analytical purposes the activity can be seen to be within a distinct system. The historical overview will provide a description of the functioning and adaptation of the decision-systems involved, before making some comparisons with the institutional response in the United Kingdom, in the next chapter, and later providing an overall transnational analysis.

THE DEVELOPMENT OF GUIDELINES

On 23 June 1976, the NIH issued guidelines to be used in all cases of recombinant DNA research funded in part or whole by the NIH.[2] The body responsible for drafting them, the Recombinant DNA Molecule Program Advisory Committee

(RAC) had spent many months since Asilomar II in producing a set of guidelines upon which they finally agreed and could recommend for acceptance by the Director NIH. In the interim period the Asilomar Summary Statement provided the only guidelines under which limited work continued.

After the momentum of the early expressions of concern culminating at Asilomar in early 1975, a wave of debate, which included many non-scientists, spread across the US. From this it had become evident that scientists within the field were not in agreement to anything like the extent that appeared to exist at Asilomar II, and in particular at the session devoted to the drafting of a statement. Discussion became vociferous at many levels including: the university campus, the city council, the state, the national level and, as discussed elsewhere, the transnational level. Within the US, however, national publicity made the deliberations at all levels of note, including the campus.

For example, many issues were aired in depth and given considerable press coverage in a debate which raged for a year at the University of Michigan. During the discussion, significant in that much of the input came from non-scientific sources, the NIH guidelines became available in draft form. The result of the extensive deliberations, involving three separate committees, was a proposal to the Regents that the NIH guidelines be followed, subject to additional safeguards.[3] In accepting the proposal, the University of Michigan demonstrated a problem for the US scientists to be repeated again in other universities, cities and states. It became apparent that the requirements facing scientists in the US might not be uniform across the land, and that local level debates, well publicised, could provide many forums in future for opposition viewpoints.

Reinforcement of these observations occurred in the internationally reported, and heated, discourse which took place in Cambridge, Massachusetts. Following a proposal by Harvard University to construct a laboratory of moderate containment, some university biologists opposed the move to allow research at that level of containment. The university debate led to the involvement of the Mayor of Cambridge, Alfred Vellucci, who announced a three-month moratorium on all recombinant DNA work involving high risk.[4] The city council then established its own committee, the Cambridge Experimentation Review Board (CERB), to investigate what safety implementation measures would be required. During the autumn of 1976, the CERB heard testimony from many scientists on both sides of the controversy before publishing, after four months, its report. In addition, many non-scientists had added pressure to curtail the work, including Friends of the Earth.[5] As a result of their deliberations, the CERB produced recommendations that the NIH guidelines be used for all work, but, like at Michigan, with a number of additional conditions to be met.[6]

By February 1977, however, further localised debates had started, at the state level in New York, California and New Jersey, and in the cities of San Diego, Madison and Bloomington (where university proposals were scrutinised).[7] Indeed, when Federal legislation was eventually proposed, the issue arose as to whether individual states should have the right to 'pre-empt' it by imposing tougher legislation. Thus, the production of guidelines in the US from the beginning would face questions of the uniformity of their application. These local debates were important elements in the wider questioning of how recombinant DNA techniques could best be controlled, and attracted some attention in the nationally centralised institutional responses, more the concern here.

Although established before the Asilomar II international meeting, the RAC had been a response to the Berg letter. Having already outlined the function required of the committee, it remains to trace its subsequent development and its efforts at drafting and later implementing guidelines. Its first meeting, on the day after the Asilomar conference adjourned, was mainly devoted to procedural business. Even so, ten members of the press joined the all-scientist group which adopted the Asilomar conference recommendations as an interim measure. Of note, the meeting also considered some early steps in the procurement of research into the potential hazards,[8] and whether there was a need for a local review function within the implementation of guidelines. However, by the time of the second meeting there had been some further developments in the US as a whole. It was during this period that the University of Michigan debate began, and in April the Senate Subcommittee on Health and Scientific Research, of the Senate Committee on Human Resources, held its first hearing on genetic engineering, under Senator Edward Kennedy.[9] In May 1975, the final draft of the Asilomar statement was submitted for approval by the NAS. The American Society for Microbiology (ASM) held a session in New York on genetic manipulation and hazards, and Harvard made its proposal for a moderate containment laboratory, again both in May.

Despite these events, which reflected a widening of the overall discussion, and the consideration, by some at least, of the risks and benefits together, the RAC would not officially follow suit. Instead, conjectured and often intuitional assessments of risk would only be set against containment requirements. At its second meeting on 12 May, the main decision of the RAC was to establish a subcommittee under D.S. Hogness, charged with beginning to draft guidelines for recommendation to the RAC. Two other decisions should also be recorded. It was proposed that a programme to develop safer hosts and vectors be set up as soon as possible, and it was suggested that a newsletter be established.[10] The newsletter would be run under the auspices of the National Institute of Allergy and Infectious Disease (NIAID) and was given the unwieldy name 'Nucleic Acid Recombinant Scientific Memorandum' (NARSM). Intended to be informal scientific communications, it is of interest in that it was an attempt to provide a link for the involved scientists, although some sections would be distributed wider. At this meeting it was also disclosed that the European Science Foundation at the time was requesting letters of understanding from those investigators it sponsored, stating that they were aware of the Asilomar principles.

After a rather pedestrian start for the RAC, it was their third meeting which introduced outright controversy into their deliberations. Often called the 'Woods Hole' meeting, it was responsible, in July 1975, for both examining and watering down the draft guidelines which were presented by the Hogness subcommittee. Confusion seems to exist, however, over the extent of agreement on this action. Only eight out of the twelve members of the RAC were present, for example. By this time, the Asilomar statement, acting as interim guidelines, was well publicised and known, but it seemed that the Hogness draft, as modified, was considerably less stringent. It did, nevertheless, give a greater description of the concept of biological containment involving enfeebled hosts and vectors, and went into more detail on the allocation of physical containment. Despite this, many critical letters were received in response to these guidelines.[11] At first the criticism came informally from those shown the guidelines, but such was the response that the RAC chairman, DeWitt Stetten, decided to circulate them more fully amongst the scientific community. A letter from Stetten to one of the minority, who actually saw the Woods Hole draft as *too strict*, is revealing. Stetten observed that:

At the conclusion of the meeting in Woods Hole, it was my impression that the membership of the Committee had indeed reached a consensus. I have since learned otherwise. I have received a flood of letters both from members of the Committee and from interested professionals like yourself. Most of these letters have argued for greater stringency.[12]

Of particular note was a letter, organised at a meeting at Cold Spring Harbor, which attracted some 50 signatures, and which expressed concern over the lowering of standards. It called for work with all mammalian DNA (not just animal viral DNA) to be categorised as requiring the second highest level of physical containment (P3; see below) and the widening of representation on the RAC. They suggested that more members expert in plant pathology and genetics and epidemiology should be appointed. Their view was that the RAC should include more scientists not directly involved in carrying out recombinant DNA work themselves.[13] Other voices of dissent were those of Paul Berg, Roy Curtiss and Stanley Falkow. Berg was willing to accept the original Hogness draft as it was before the revisions of the RAC, while Falkow's objections arose out of his being one of the members of the RAC who missed the meeting. Curtiss, however, wrote at length criticising the likely success of biological containment, with supporting evidence from work in his own laboratory, which had been trying to produce modified *E. coli* through mutation selections. He also questioned the idea of enabling trade-offs between relaxing physical containment in exchange for higher biological containment. In sum, he saw the Woods Hole guidelines as 'contrary to the spirit of the Asilomar meeting'.[14]

A member of the Boston Area Science for the People group, Jonathan King, a biologist at MIT, was particularly critical of the fact that four members of the Hogness subcommittee were involved in developing recombinant DNA techniques. King added to calls for wider representation on the RAC, suggesting that it should include someone from the Environmental Protection Agency (EPA) or the Occupational Health and Safety Administration (OHSA). A detailed and technical critique of the guidelines was produced by King's group and sent to the NIH, but 'was never published since no avenue of publication was available for such a document'.[15] Indeed, the future activities of the RAC, even beyond the completion of the guidelines, would be influenced and monitored by a rising tide of interested bodies. In essence, the disagreements reflect the political nature of the decision-making exercise. Critics of the process had then already raised the important issue of participation. King, however, felt the pressures of counteraction against this. As he claimed: 'The critics were continuously referred to pejoratively by the proponents as "kooks", "Those who have inflamed the public ..." "incompetents", "those who want to destroy science". Life became quite unpleasant for those of us who were trying to bring the issue out into the open.'[16]

Hogness defended the actions of his subcommittee and the RAC as a whole on the production of guidelines. His view was that the critics were not weighting the benefits of research, in their widest sense, against the conceivable risks.[17] It could be said of this, however, that scientists always assumed benefits, even though they would occur well into the future. The controversy that arose as far as the guidelines were involved was not over the definition of the containment levels, but over the allocation of types of experiments to those categories. Subsequent meetings of the RAC would return to this central problem time and time again. An example of the debate concerned the containment requirements for 'shotgun experiments' using genetic components of animals closest to man, such as other primates. Under the Woods Hole proposals the highest containment level was not prescribed.

As a result of the criticisms of the Woods Hole draft, the RAC chairman appointed a second subcommittee to produce a tighter set of guidelines. Elizabeth Kutter, a supposed 'lay' member of the RAC, appointed at its request, was to chair the subcommittee. As a biophysicist (although she worked with phages) accusations of self-serving might be avoided. The NIH view had, therefore, responded, if only to a limited extent, to questions about the legitimacy of the RAC structure.

Prior to the fourth meeting, a workshop was held, sponsored by NIAID, on the 'design and testing of safer prokaryotic vehicles and bacterial hosts for research on recombinant DNA molecules'. Some 60 scientists attended the meeting organised by, among others, Curtiss and Falkow. A critical requirement, it had been recognised at Asilomar II, was the development of appropriate enfeebled bacteria. The excessive optimism that a suitable strain would be developed in weeks had been somewhat modified by the time of this workshop. As a measure of the problem, it was the lack of such an organism that led to some experiments having been under a moratorium for some 18 months.[18] The Woods Hole meeting had been handicapped by the absence of these enfeebled strains, in turn reinforcing the RAC objective of developing safer hosts and vectors.

Between 4 and 5 December, the RAC held its fourth meeting, starting the day after the NIAID workshop and in the same town. A summary of the workshop was presented by Dr Helsinki. He reported the good news that suitable plasmids and lambda phage vectors and hosts had now been developed, and were awaiting testing. Ironically, despite his earlier misgivings, the breakthrough had occurred in Curtiss' laboratory.[19] Applicable to all three drafts so far produced, the Hogness, Woods Hole and Kutter versions, the use of enfeebled organisms could now be written in with some confidence. Thus the fourth meeting held at La Jolla, California, set about reconsidering, in effect, the business of the previous meeting. In addition to the now 15 members of the RAC present, there were 42 others. Eight were classed as '*ad hoc* consultants to the committee', three were 'liaison representatives from other organisations' (including the NAS and NSF), eight were NIH staff, five were members of the press and three were from Europe, including Sydney Brenner of the UK Medical Research Council and John Tooze of EMBO. Most of the others were scientists from various laboratories.

There was a general feeling that this particular meeting must produce draft regulations or 'the dam would be likely to break'.[20] Unconfirmed rumours were circulating concerning experiments being carried out clandestinely as patience began to wane. The relevant US scientific community was directing pressure for the RAC to produce something, while at the same time the eyes of other states were following developments closely. Some were delaying the production of their own controls as they awaited the US outcome. In the event, guidelines more strict than those of the third meeting were forthcoming, perhaps as a result of two main factors: first, there was the relatively imminent provision of suitable enfeebled host–vector systems (of EK2 level, see below); secondly, three of the most influential organisers of Asilomar II were present: Berg, Singer and Brenner. The existence of the potential for biological containment bypassed pressure for relaxing that provision, as had occurred at Woods Hole, and the Asilomar organisers added their influence in support. Thus the 'weaker-than-Asilomar' proposals of the previous meeting gave way to a 'stricter-than-Asilomar' new set.[21] A background threat, perceived by those involved, assisted their motivation. If they failed to act, then others might take over, and in particular that might mean Congress. It would be interesting to speculate as to what the outcome might have been had the enfeebled organisms not come along at that time. Adoption of weaker guidelines might have precipitated the very consequences the scientists feared – greater public controversy and legislation.

As an individual, Sydney Brenner has been recognised by commentators of the time as influential in the maintenance of caution. Repeating his performance of Asilomar II, he applied his powers of reasoning and his oratorical skills to instil a sobering influence. Caution was implemented in a novel fashion. By the time of the fourth meeting, there were three drafts to consider, those of Hogness, Woods Hole and Kutter. Ground-rules were established to enable a paragraph-by-paragraph comparison. No more than ten minutes would be allocated to each paragraph in the first instance, and were no discussion developed then the chairman would simply choose a version. Difficult points, needing more than the allotted time, would be returned to, and although only the committee could vote, the chair would recognise all in attendance if they wished to speak. Many paragraphs were passed over without discussion, while some required a simple vote on which of the three versions would be chosen. Only with paragraphs or sections which needed complete or partial redrafting was there any difficulty.

An example of the debate can be drawn from the consideration of the 'shotgun experiment'. Hogness argued that *particular* experiments (such as any using DNA from cold-blooded vertebrates) should have the risk explained in detail. Brenner, however, responded by suggesting that the class of source DNA was not so relevant, in the case of shotgun experiments, as the type of experiment *per se*. The risk lay with fragmenting such large samples of genome, whatever its source. Brenner also warned the RAC of outsiders viewing their actions in terms of trade-offs and bargains over the assignment of particular categories to DNA of various classes of source.[22] Indeed, future criticisms of the final guidelines in both the US and the UK would in part be directed against the rationale of assigning risk in relation to the biological distance of the donor organism from man.

Finally a draft was completed and presented to the Director NIH for his approval. However, before this was forthcoming, the Director would hold further discussions, including a public debate. Meanwhile, at La Jolla, the RAC considered further business. It was hoped to receive all contract proposals for the construction and testing of safer hosts by January 1976, and to offer contracts by March. It was decided that the RAC, for the time being, would certify proposed host–vector systems, and would give advice regarding the usage of clones constructed under the interim Asilomar recommendations. Of note, though, was that the chairman was asked if the NIH would explore what impact the RAC would have on industrial applications of recombinant DNA technology. And Brenner suggested that an experiment be carried out to test whether biological mechanisms existed to transfer DNA from a lower organism to a higher organism. Under maximum containment, a polyoma virus would be inserted into a plasmid, in turn put in a bacterial host and placed in new-born mice. The mice would then be monitored for infection by the virus. Unanimously, the committee decided to request NIH assistance in developing such an experiment. Lastly, the RAC set up a subcommittee to design further experiments aimed at assessing biohazards.

Undeniably, the RAC was a central actor in the US institutional decision-system and had great influence with the Director NIH, Donald S. Fredrickson. It was not, however, his only source of advice. Early in February 1976, he called a special meeting of his 'Advisory Committee to the Director' to review the proposed guidelines. This committee was charged with advising the Director on matters relating to the broad setting within which the biomedical sciences developed, including scientific, technological and socioeconomic factors. Membership included individuals knowledgeable in the fields of basic science, clinical biomedical sciences, physical sciences, social science, research, education and communications.[23] For this particular meeting, additional participants were invited, for example past members and other scientific and public representatives. Seven people made presentations, including Fredrickson, Stetten, Singer, Hogness, Berg and Curtiss. Thirteen members of the public made statements, and a further 19 individuals attended. A wide range of views was presented, a credit to Fredrickson's notification of the meeting to a large number of public interest-groups.[24] Such was the range of backgrounds at the meeting that Michael Rogers has commented: 'Compared to the nearly exclusive academic background of the guidelines committee this was really almost the public.'[25] There was much more public interest than expected and more press in attendance than for any previous meeting. Singer and Berg outlined the guidelines, but as Rogers observed, the real contribution of the meeting was that it provided a forum for the critics to express their views. Most critics were young and had little scientific background, yet were critical of the motives and possible gains of the original guideline writers. Rogers agreed, however, that at least some bias was evident in the selection of magazine articles given to the Advisory Committee to read, as they tended to defend the guidelines. They were not representative of the range published in journals such as *Science*.

Overall in the recombinant DNA debate, and particularly around this time, two critics were of note: Robert Sinsheimer, a member of the Advisory Committee who participated in the February meeting, and Erwin Chargaff. Sinsheimer, Chairman of the Division of Biology at the California Institute of Biology, had, for example, sent a letter to Fredrickson prior to the meeting,[26] which raised four key points. First, he saw the failure of physical containment as inevitable in the long run, given the likely number of laboratories which would move into the field.[27] That is, physical containment was limited and dependent on human efficiency. Secondly, Sinsheimer questioned the effectiveness of biological containment in that genetic information might be passed from enfeebled strains of organisms used in the work to other organisms through natural recombination. Man himself contained many organisms capable of initiating and carrying out such recombination with *E. coli*. Thirdly, Sinsheimer raised the possibility of 'breaking' the genetic species 'barrier' between prokaryotes and eukaryotes through the transferring of DNA segments with their promoter or terminator sequences, thought to differ on each side of the 'barrier'.[28] Thus he warned of upsetting 'an extremely intricate ecological inter-action which we understand only dimly'. Finally, Sinsheimer introduced a point he felt was underestimated, namely that they were dealing with self-replicating organisms potentially making any hazard, should it appear, irreversible. This situation, he inferred, was unlike any other physical or chemical hazard.

As a consequence of the points he raised, Sinsheimer recommended to Fredrickson that all recombinant DNA work should only proceed under the highest possible precautions and at one national site. This most eminent biologist went as far as to say that he would only allow this much because of the extraordinary benefits that could accrue. He concluded: 'I will say, though, that in my judgement, if the guidelines are adopted and nothing untoward happens, we will owe this success far more to good fortune than to human wisdom.'[29] After the meeting of the Advisory Committee, Sinsheimer repeated his concerns in another letter to Fredrickson.[30] In this he challenged some explicit defences addressed to the critics, which had been made at the meeting. It had been argued, for example, that a justification for proceeding with the research was the fact that recombination occurred in nature. Sinsheimer, however, argued that this occurred at some unknown frequency which man would be increasing by 'many orders of magnitude'. He still feared irreversible consequences, whatever their probability, of man-made novel gene combinations entering the gene pool of the environment. In addition to his earlier comments, Sinsheimer suggested a way to proceed with the work which in general would be unusual in basic biological research outside industry. He suggested that the starting-point of research should be the *objective*. If synthesis of a particular protein could be deemed desirable, then safe procedures could be tailored for that objective, rather than trying a multifunctional approach. Further, he repeated worries about using *E. coli* by arguing that a suitable animal virus should be found, to which we have known defences, rather than using 'free-living ubiquitous bacteria'.

Yet, despite his views, Sinsheimer agreed with that part of the mood of the Advisory Committee meeting that seemed to fear future legislation and, in particular, any requirement for prior approval of experiments, which, he thought, might impede science. Whatever misgivings Sinsheimer had on this latter point it is worth noting that it need not have been too problematic as prior approval was precisely the approach used in the UK to commendable effect (despite some discontent from certain quarters). Erwin Chargaff, Professor Emeritus of Biochemistry at Columbia University, was the second of the two influential critics who produced forceful and somewhat authoritative arguments. Their standing makes them of note, whatever the judgement that can be applied to their arguments in retrospect. Not in attendance at the Advisory Committee meeting, Chargaff did, however, write to Fredrickson on the day before. He called for a two-year 'cooling-off' period, involving prohibition of the work, accompanied by 'discussion and reflection'. He stated: 'I can only express my strong dismay about the whole program as it has been formulated, restricted, expanded, and drowned in a verbiage of silly claims and counterclaims.' Soon afterwards he wrote to *Science* commenting, like Sinsheimer, although rather sarcastically, on the undesirability of using *E. coli* and the potentially irreversible ecological significances.[31] However, of particular note was that he broached the ethical question in claiming that the issue was not one of public health, but was more related to their right to put 'an additional fearful load on generations that are not yet born'. The NIH, he suggested, was not the appropriate institution to address such problems.

These arguments of Sinsheimer and Chargaff have been presented at some length, partly because of the importance of the points made given the knowledge of the time, and partly because opposition groups rallied to them. Non-scientists, in particular, tended to identify with scientists reflecting their views in order to enhance the legitimacy of their own arguments. The two critics in short were preaching caution and social responsibility. It should be stated, nevertheless, that the arguments concerning the crossing of the 'barrier' between eukaryotes and prokaryotes were not accepted by many scientists. One group had, for example, argued that any such barrier might have no purpose and simply reflect different evolutionary paths. A second group had argued that it did not exist at all and that they were merely ignorant of the flow of genes between species. A third group suggested that evolution had already tested all genetic combinations and dismissed those not useful. Sinsheimer attributed many such arguments to those who emphasised the benefits to be gained from the research. For example, he pointed to a flaw in the third group's argument. If nature had tested all gene combinations, then their own proposed genetic manipulations could not be novel, and by implication beneficial.[32] To an extent it was the existence of this division of views between respected scientists that is of interest here, as it reflects the latitudes involved as a result of uncertainty.

Numerous others also wrote to Fredrickson, including Berg, Singer, Handler and Goldstein. Berg and Singer expressed general satisfaction, although Singer did suggest greater stringency in dealing with SV40. Goldstein urged a meeting of both sides to avoid 'opting only for development, rather than development, leadership and control of a new technology'.[33] A wide variety of issues were therefore aired after the drafting of guidelines at La Jolla. One more RAC meeting would occur before final publication, and for this Fredrickson recommended discussion should centre on a document which he distributed. It was a summary of the responses which he had received after the February meeting of his Advisory Committee.[34] Under a number of headings Fredrickson provided a summary of comments which he had received, whether technical or more general. These were: Methods of Containment; Prohibited Experiments; Biological Containment Criteria...; Classification of Experiments ... [using the new enfeebled *E. coli*]; Experiments with Eukaryotic Host–Vectors; and Implementation. It was, however, noticeable that Fredrickson tended to favour certain views which were for the most part those less critical of the draft guidelines. For example, despite comments from some scientists going so far as to suggest a complete ban on SV40, he made it quite clear that he favoured its use, and under the provisions of the planned guidelines. The difficulty with the course of action chosen by Fredrickson is that it can be criticised on the basis that it was the role of the RAC to present him with advice and not the other way round.

At the RAC meeting held in early April 1976, most of the suggested amendments were rejected, and in reflection of the tone of Fredrickson's document broader issues, as raised by Sinsheimer and Chargaff, were not discussed. Nicholas Wade reported the outcome as 'a clear victory in sight for those who wish research to go ahead under stiff but not grossly inconvenient safety conditions'.[35] If the suggestions of Fredrickson guided the RAC meeting, then it should be noted that Fredrickson himself faced constraints. Wade identified these as: first, an effective veto held by the European Molecular Biology Organisation (EMBO) which had made it clear that it would only go along with the NIH guidelines if they became no stricter; secondly, Wade suggested that there was a fear that if the guidelines were unacceptably rigorous, then the scientists would ignore them! An assiduous follower of the events as they unfolded, Wade identified two members of the RAC as dominant, David Hogness and Charles Thomas, of Stanford and Harvard, respectively. Both were forcefully against stricter guidelines and both were personally involved with recombinant DNA work. Whether or not they genuinely reflected a majority view of the scientists at large, they were open to the charge of conflicts of interest.

In the interim period between the fifth RAC meeting and publication of the guidelines in June, the Director, NIH received further correspondence.[36] Berg, for example, expressed dissatisfaction with biological containment preferring emphasis on physical containment as the 'only tried, tested and acceptable means available for containing dangerous pathogens'. He also noted that in the UK it was physical containment on its own which was being relied upon. A non-scientist, J.F. Kelly, requested that a broad policy statement be attached to the guidelines. He argued that as it was, they were in effect a set of implementation procedures, the underlying policy of which had to be inferred. Others raised questions of the appropriate actions if environmental contamination were to occur, and the assumptions in general upon which the NIH and RAC were operating. In technological fields, preventive measures are often taken, but without contingency plans for failures, perhaps because the latter could imply a weakness in the safeguards. The Federation of American Scientists despatched to the NIH the results of a poll of their members. Their results showed only a small percentage of biologists (9.3 per cent) and non-biologists (5.7 per cent) who thought the guidelines were too restrictive. Overall, more than 80 per cent of both biologists and non-biologists who responded thought the guidelines either about right or 'probably insufficiently cautious'. In all, some 56.2 per cent thought the guidelines 'about right'.[37]

An even more systematic poll was later organised by a science correspondent for the *Boston Globe*. Robert Cooke, between February and March 1977, mailed questionnaires to some 1250 senior biologists working in universities, industry and government. Of the 490 respondents, 39 per cent believed the guidelines should be more strict, while 44 per cent thought them strong enough.[38] Both of these polls reflect the degree to which caution was seen as desirable by many interested scientists and non-scientists. It is of note that the RAC itself was unanimous in its recommendation to Fredrickson to go ahead and publish. Yet as late as a few days before publication, a petition carrying some 30 signatures called for a postponement on the release of guidelines.[39]

Having considered the many comments received over the months, Fredrickson, on 23 June 1976, released the official NIH guidelines, some two years after the publication of the Berg letter. They were accompanied by a document drafted by Fredrickson which summarised the history of the debate, thus far, and indicated his own attitudes.[40] To be fair, he provided a useful synopsis of the views discussed above and in the many representations he received. Apart from making some general comments regarding the speculative nature of hazards and the by then more certain promise of benefits, Fredrickson failed to detail the assumptions and policy choices to which he was working. At one point, for example, he summarised the two arguments concerning the 'barriers' between eukaryotes and prokaryotes. On the one hand, recombination in nature was argued to have been occurring for so long that man's efforts would merely utilise specific combinations of genes. On the other hand, man could cross the barrier in the laboratory and if the donor DNA was expressed in the host, it might alter the host in 'unpredictable and undesirable ways'. Fredrickson concluded: 'The fact is that we do not know which of the above-stated propositions is correct.'[41] Nowhere does Fredrickson give any indication of the strength of feeling that was, for example, expressed by Sinsheimer and Chargaff over potential environmental damage and the evolutionary effects. A further example of his style of argument which is of some concern can be seen in his resolution of the fears expressed over the use of *E. coli*, which in one of its natural forms inhabits the human gut. Fredrickson countered this by arguing that because the bacterium was also the most understood (the laboratory strain which would be used) it should therefore be 'safer than other candidate micro-organisms'. Any underlying method of balancing such arguments is not revealed, and in effect the guidelines are much as recommended by the RAC.

Perhaps to avoid charges of self-serving, as after all the NIH had a responsibility to *promote* research, some assumptions influencing decisions and policy were not made too explicit. In organisational terms this would imply bias and non-decision.

THE FIRST US GUIDELINES FOR RESEARCH INVOLVING RECOMBINANT DNA MOLECULES

Four principles were stated as underlying the published guidelines:

(i)   There were certain experiments for which the assessed hazard was so serious that they were not to be attempted at that time.

(ii)   The remainder could be undertaken with appropriate safeguards of a physical and biological kind.

(iii) The level of containment should match the estimated potential hazard of the different classes of recombinants.

(iv) The guidelines themselves would be subject to periodic review (at least annually).

Four levels of physical containment were incorporated in the guidelines, ranging from P1 to P4 as the physical design of the laboratory, the equipment and procedures used increased in complexity. P1 would be in essence an ordinary microbiological laboratory where work could be carried out on open bench tops. At the other extreme, P4 containment would involve a whole host of measures to limit the potential for environmental contamination. The laboratory at this level should be effectively isolated, maintained at negative air temperature, be completely sealed as a monolithic unit except for stringently controlled access, and exhausted air should be decontaminated. Sophisticated double-door autoclaves would be used for sterilisation and work would be carried out in safety cabinets. With these and other features the laboratory would be a maximum security unit of the likes of Fort Detrick in the US and Porton Down in the UK. In summary, the physical containment was to cover minimal (P1), low (P2), moderate (P3) and high (P4) risk levels.

DNA from non-pathogenic prokaryotes that naturally exchange genes with *E. coli*	P1 and EK1
DNA from embryonic or germ-line cells of cold-blooded vertebrates	P2 and EK1
DNA from non-embryonic cold-blooded vertebrates	P2 and EK2
DNA from non-pathogenic prokaryotes that do not naturally exchange genes with *E. coli*	P3 and EK1
DNA from embryonic primate-tissue or germ-line cells	P3 and EK2
DNA from non-embryonic primate tissue	P3 and EK3 or P4 and EK2

3.1 Containment requirements for a selection of experiment categories from the 1976 US Guidelines

The guidelines also designated three categories of biological containment, EK1, EK2 and EK3. EK1 containment referred to the use of the basic laboratory strain, *E. coli* K-12 with existing plasmid and bacteriophage vectors. Apparently a harmless microorganism, it does not usually establish itself in the bowel, but if ingested it stays alive in passing through. However, allowing for the possibility of genetic exchange between *E. coli* K-12 and other residents of the gut, the organism was thought to offer only moderate containment. EK2 containment would utilise genetically constructed host–vector systems demonstrated to provide a high level of biological containment through data from suitable laboratory tests. Modifications to the *E. coli* K-12 host or the plasmid or phage vectors were to be such that a genetic marker, carried on the vector, should not show up in other than specially designed and carefully regulated laboratory environments at a frequency greater than 10^{-8} (or one in 100 million). EK3 containment was designated as the same as EK2, but with the specified level of containment shown to exist through appropriate tests in animals, including humans and primates, and other relevant environments. This additional validatory data would set EK3 apart from EK2 as a category, by increasing certainty.

A number of experiments were banned, including any involving the use of DNA from dangerous pathogens,[42] or oncogenic viruses, or DNA from cells infected with such agents. Also banned was the use of DNA containing the genes coding for the biosynthesis of potent toxins such as botulinus and diphtheria, or the venom from snakes. The use of plant pathogens was similarly excluded. In addition, certain other activities, of less direct hazard, were to be avoided at that time, including the transference of a drug resistance trait to an organism not known to acquire it naturally, the release of any recombinant DNA molecule into the environment, and finally, the exceeding of the maximum of 10 litres of culture to be used in recombinant DNA work.[43]

All the remaining experiments were identified in categories related to the biological distance from man of the DNA used. Each category was then assigned physical and biological containment requirements, and in some instances alternative combinations of levels were permitted. For example, P3 and EK1 was considered comparable with P2 and EK2, while P4 and EK2 could substitute for P3 and EK3. Well summarised by a number of commentators,[44] some examples of the containment classifications are shown in Figure 3.1.

In addition to biological distance from man, containment was also applied on the principle that containment should never be less than that already required of the most hazardous component of the experiment in existing guidelines or regulations. However, the guidelines suggested that a further precaution be reflected in the undertaking of work involving recombinant DNA techniques. It was emphasised that there was a need to ensure thorough training in microbiological practice, including aseptic techniques and instruction in the biology of the organisms used in experiments so that potential biohazards could be understood and appreciated. Laboratories were required to prepare emergency contingency plans and to make it known to workers where experiments involved 'known or potential biohazards'.[45]

One of the main features of the NIH guidelines was their comprehensiveness. This arose out of the policy adopted in the US to issue guidelines from a central source, but leave the identification of containment categories regarding individual experiments with the laboratory and institution concerned. The UK system will later be contrasted, in that allocation of containment levels for all proposed experiments was centrally determined. A large proportion of the NIH guidelines was therefore given over to their implementation.

APPLYING THE GUIDELINES

For the moment, discussion of the implementation of the NIH guidelines will refer exclusively to their use in the US. It is, however, recognised that their influence was international, and as a prelude to outlining their international impact (in Chapter 5), it is worth recording that in June 1976, the State Department despatched a telegram to 44 of its embassies and missions in countries believed to support considerable bio-medical research. It requested that appropriate officials of host government agencies be informed of the impending release of guidelines and the address of the Director, NIH. The telegram, not least, acknowledged the need for 'world publicity and co-operation on the problem'.[46] It was the comprehensive categorisation of types of experiments in relation to containment which made the NIH document an attractive 'off the shelf' option for other countries. Despite this, its implementation within the US itself was not without difficulty.

An issue of particular importance was the enforcement of the guidelines. The NIH had opted to avoid elevating the guidelines to Federal regulations. Acknowledging that many commentators would prefer this course, Fredrickson stated that the scientific community in general argued against it. He added his support to these scientists, suggesting that there would be more flexibility and administrative efficiency if Federal regulation was avoided, although he acknowledged that the whole matter needed further attention. The question of ensuring compliance with the guidelines had, however, a further weakness. As they stood, the guidelines only applied to

NIH-funded research, of which the control of grants could be called upon to give them credibility. Yet research was always likely to be done by many who were not funded, even in part, from the NIH.

At the time of the release of the guidelines, first moves were in hand to have inter-Federal agency discussions to assess the possibility of widening their application. This activity reflected one strand of the problem, a potential solution being other agencies requesting compliance as a condition of funding. A more disturbing problem to many at the time was the question of ensuring compliance from wholly industrial sponsored research. For his part, Fredrickson, prior to the release of the guidelines, had held a meeting with industrial representatives, under the auspices of the Pharmaceutical Manufacturers' Association (PMA). Fredrickson had been hoping for the acceptance of the guidelines throughout the US, and, at least similar guidelines, internationally.[47] Although Fredrickson's statement accompanying the guidelines is not very revealing regarding the attitudes of industry at that time, some letters which he subsequently received do throw some more light on their views.

W.N. Hubbard of the Upjohn Company gave two reasons why industry felt concerned: first, he expressed a fear, often to be repeated internationally, that industrial confidentiality could be compromised if research results were disclosed prior to patenting; secondly, Hubbard stressed a worry about industry becoming involved with self-serving regulations. He concluded:

It is my impression then that industry will avoid committing itself in very formal ways to policy statements out of a reasonable and deep-seated fear that this is just the first step of another wave of bureaucratic intervention into individual endeavor.[48]

C.W. Pettinga of Eli Lilly and Company indicated that his company would 'adhere to the intent and spirit of the NIH guidelines' and had taken 'several steps towards this end'. A safety committee had been set up and a P3 laboratory established. However, Pettinga suggested that industry had more experience than anyone else in working with large-scale contained studies, and that if they were convinced of the safety, they would have no hesitation in exceeding the ten litres maximum level. Further, the intent to follow the spirit of the guidelines was qualified by saying:

There are specific instances where we might not be in strict compliance. In these specific instances we feel either that the wording of the guidelines is too non-specific or we are capable of guaranteeing the safety of the exercise in question.[49]

Interestingly, Pettinga also suggested that the RAC membership should include a well-informed representative of industry.

Two pressures were being experienced at the time. On the one hand, industry wanted to wait and see what the guidelines would be like and the extent of their enforcement before becoming too public in their opinions. On the other hand, however, they faced competitive pressures to actually get ahead and utilise the new research tools. Conflicting pressures of this sort would not be conducive to effective operational control of a voluntary nature. A telling summary of the meeting in June, which 30 industrial representatives attended, observed that:

In general, the industries seemed to be somewhat hesitant to commit themselves with regard to the guidelines since it was believed that the guidelines might eventually assume the status of regulatory law and this would place an entirely different perspective on their views about the details of the guidelines.[50]

This summary went on to suggest that although the President of the PMA had implied during Congressional hearings that the guidelines were endorsed by the drug industry, this spirit was general through all industries.

Thus, the NIH guidelines at their inception were faced with a problem of ensuring compliance in their implementation and the treatment of industry in particular was to remain a somewhat controversial issue. This is given further attention below. Nevertheless, the 1976 guidelines did describe roles and responsibilities for persons and organisations concerned with recombinant DNA activity. From this an outline of the operational control system can be drawn, as illustrated in

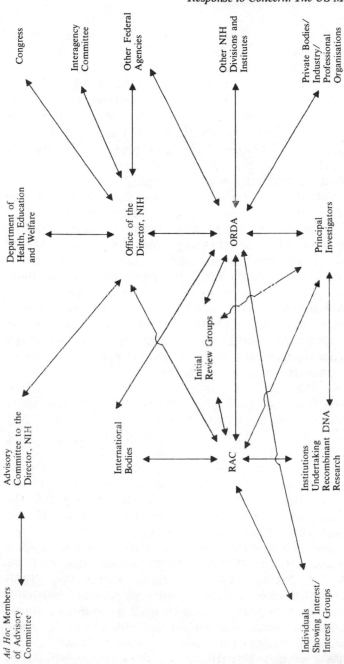

3.2 *Diagrammatic illustration of the central roles of the RAC and ORDA in the United States, response to the recombinant DNA debate.*

Figure 3.2. A quite complex network was involved at the centre of which were two bodies in particular; the RAC and the Office of Recombinant DNA Activities (ORDA). The guidelines announced the existence of ORDA which was to fulfil a number of roles including administratively servicing the RAC, collating information, and communicating information to the many concerned groups. It was to have close ties with the Office of the Director, NIH, take part in planning and develop relations with other Federal Agencies, the science press, industry and professional groups.[51]

Responsibilities of other groups were also described in the guidelines. NIH staff assessing grant applications were to restrict awards to those who formally completed a Memorandum of Understanding and Agreement (MUA) which acknowledged responsibilities, assessed containment and described laboratory facilities. They were to undertake inspections of P4 facilities, assess requests to lower containment on rigorously tested products of shotgun experiments and respond to enquiries. NIH Initial Review Groups would review the scientific merits of each grant application and make independent judgements on hazards and proposed containment safeguards. However, a great deal of responsibility was to fall at the level of the research institution and the researchers involved. Each laboratory proposing to undertake recombinant DNA work would have to appoint a Principal Investigator, who would have responsibility for a whole host of functions from estimating the hazards involved to determining the appropriate containment and training their staff. Not least they would maintain contact with the Institutional Biohazards Committee (below) and ORDA, offering any information pertinent to the guidelines. If they wished to step outside the strict conditions of the guidelines they would need approval from the RAC. Each institution allowing recombinant DNA research would also have to establish an Institutional Biohazards Committee (IBC) charged with: advising the institutions on policies; creating a central reference of relevant information; developing a safety and operations manual for any P4 facilities; certifying to the NIH on applications for research support that all the conditions were met. The IBC was to comprise individuals of 'a diversity of disciplines relevant to recombinant DNA technology, biological safety, and engineering'. In addition, it was to possess, or have available to it, the competence to determine how its findings related to applicable laws, regulations, standards of practice, community attitudes and environmental considerations. Indeed, the last point would help to enhance the legitimacy of the committee within its local area, and its legitimacy as an overall component of the operationalisation of the guidelines. To supplement this, minutes of the meetings were to be made public.

The Office of the Director, NIH, was to be responsible for the promulgation and enforcement of regulations or guidelines, and for accountability to Congressional committees, the Department of Health, Education and Welfare and the public.[52] However, the RAC would have the duty of future guideline revisions. It would also certify EK1 and EK2 systems, resolve issues of containment if requested by NIH staff and, with an eye towards large-scale work, review and approve experiments involving over 10 litres of culture.

Thus a framework for US research was formalised, even though the assignment of roles and responsibilities had not been without criticism during the planning stages. Fredrickson, himself, in his statement accompanying the guidelines, recognised the conflict of opinions. With regard to IBCs, one view had been that they should be *required* to determine containment conditions for given projects. (In the draft guidelines it had been stated explicitly that this function would not be carried out by IBCs.) The RAC opposed the suggestion, arguing instead that the most appropriate level for scrutiny of containment conditions was the national level, using NIH study sections. Fredrickson again applied his own judgement and chose to take a line whereby IBCs would not be required to undertake such a function, but could if their institutions so wished it.

The structure of the RAC had also involved some differences of viewpoint. For example, it had been suggested that the scientific advisers on the committee should include a number not involved in undertaking recombinant DNA research, and that in addition there should be a committee to offer more policy-oriented advice. Fredrickson opted to defend the inclusion of involved scientists in that they would have 'the expertise to assure that the guidelines are of the highest scientific quality' and would in any case be complemented with scientists from other fields.[53] On the policy side, Fredrickson noted that the RAC itself requested the inclusion of a non-scientist, and that a Professor of Government and Public Affairs had duly been appointed. An ethicist had also been nominated. However, the main source of policy advice was to be the Advisory Committee to the Director, considered above. The two-committee structure, commendable in that it provided a wide range of expertise and political advice, was weaker than necessary in that both committees independently advised the Director. There was little opportunity for cross-fertilisation of ideas between the committees themselves. In the United Kingdom, for example, a single central committee comprised many scientists and non-scientists and could address a wider range of issues than the RAC. Thus, in the last resort, the Director, NIH, himself had considerable latitude in how he synthesised advice and along with the RAC, and in the future ORDA, he fulfilled a central position in the formal framework.

A number of criticisms have, therefore, been presented in passing, and further analysis will be produced after the examination of activities in other states, notably the UK. However, within the US, the Director, NIH, himself was legally required to make an analytical appraisal of his actions in establishing guidelines, in the context of an Environmental Impact Statement.

ASSESSING ENVIRONMENTAL IMPACT

Early in 1976, the Director, NIH began to receive requests, some in the form of threatened legal action, for him to publish an Environmental Impact Statement (EIS) on proposed actions relating to recombinant DNA. The National Environmental Policy Act (NEPA) of 1969 was cited as applying to all agencies of the Federal Government, covering 'every recommendation or report on proposals for legislation and other Federal actions significantly affecting the quality of the human environment'. Under this law, a detailed statement was required on: environmental impacts of the proposed action and environmental effects which could not be avoided; alternatives to the proposed action; long-term implications; and any irretrievable use of resources involved.[54] In complying, Fredrickson opted to produce an EIS as soon as possible, but not at the expense of holding up publication of the guidelines. His view was that the whole process of developing guidelines was 'in large part tantamount to conducting an environmental impact assessment'.[55] It was clear, however, that many thought the analysis involved was narrowly based and inadequate.

Thus it was on 9 September 1976, some eleven weeks *after* the guidelines were released, that a draft EIS on their impact was published in the *Federal Register*.[56] The public were then invited to submit comments by 18 October 1976, and one year later the final version appeared which incorporated the comments received.[57] In a very descriptive fashion, the EIS summarised the main events, underlying principles and assumptions leading to the guidelines. Of particular importance was that it was required to address different possible courses of action with respect to the new research techniques. Five options had therefore been identified in the draft EIS:

(i)   No action.
(ii) NIH prohibition on the funding of all experiments with recombinant DNA.
(iii) Development of different guidelines.
(iv) No guidelines, but NIH consideration of each individual proposed project before funding.
(v)   General Federal regulation of all such research.

The 'no action' choice on the part of the NIH would have left researchers with only the Asilomar Statement to act as guidelines. Public concern would have been higher, while the costs to researchers in both time and resource terms would have been less. Aside from the public concern, such an option would, in effect, have meant ignoring the explicit call for NIH involvement expressed in the Berg letter.

A refusal by the NIH to fund any recombinant DNA research would not imply a complete US ban, as other public and private sources of funding existed. Not least, industry would have still shown interest. From the NIH point of view, two main arguments were used to dismiss the option. First, an intuitive risk-benefit argument was used to emphasise that many benefits were to be gained from the research, and therefore it should continue. This does not overcome a possible criticism that a more thorough risk-benefit assessment should have been undertaken before allowing the research to proceed. The second argument was more worrying. It was illogical and irrelevant as far as any safety considerations were concerned. In essence, having acknowledged how US scientists had played a major part in drawing attention to the potential hazards, the draft EIS argued that the NIH guidelines were likely to be accepted as a model internationally, and that 'prohibition of the work would undermine American leadership in the establishment of worldwide standards for safety'.[58] This can easily be countered by suggesting that US caution to the extent of extending the moratorium might also provide the international lead regarding safety. In terms of prestige, however, it could be suggested that the harsher the restrictions faced by US researchers in an internationally competitive field, the less likely it would be that other states would in fact follow suit. A more honest argument related to this, reproduced in the draft EIS, was that the banning of the work in the US could undermine its leadership in biological research, if the work continued elsewhere. In fact the whole associated question of international harmonisation in guidelines and safeguards is something to return to.

Discussion in the EIS on 'different' guidelines was confined to giving examples of controversial differences of opinion over the containment requirements for specific experiments. It has already been shown how arguments differed in this respect. More fundamentally, the NIH can be criticised for not justifying more thoroughly the underlying principles and assumptions it operated upon. For example, it never really explained why immediate health concerns should dominate over ethical or long-term ecological considerations.

The possibility of the NIH assessing each individual

proposed experiment, in relation to criteria applied by a panel or committee, was considered by way of an alternative to the existing guidelines. Whatever the value of making such a comparison after the guidelines option had been chosen, the draft EIS quickly dismissed the possibility. Its advantages were recognised as providing a system of greater flexibility in adapting to new knowledge. Disadvantages in a US context would, however, be notable, even though this was essentially the approach adopted in the UK. Enormous time and resource costs would apply given the size of the US and the potential research interest. It was doubted whether sufficiently knowledgeable individuals could be found for what were likely to be full-time jobs on the central committee. Besides, it was argued that IBCs would initially assess each proposal, which would then be re-evaluated by the NIH Study Section in reviewing the scientific merit of the proposal. At best, this would be seen as a compromise. More will be said of central experiment safety assessment when examining the UK institutional responses.

Perhaps the only option given serious consideration as an alternative to the NIH implemented guidelines was the possibility of Federal regulation of all recombinant DNA work. This option was, in particular, given substantial coverage in the final EIS as a consequence of a development between it and the earlier draft EIS. In October 1976, the Secretary of Health, Education and Welfare had established, with approval from the US President, an Interagency Committee on Recombinant DNA Research, chaired by the Director, NIH. This committee was important as Fredrickson had felt that the question of Federal regulation was beyond the purview of the NIH, a research agency. It was intended that the Interagency Committee would determine the applicability of the NIH guidelines to industry, and to other Federal agencies. All Federal departments and agencies which might support or conduct relevant research, and all regulatory agencies which might have potential authority over it, were represented on the committee. If necessary, the committee was to recommend appropriate legislation or executive action. Nineteen different bodies were represented, including the Department of Defense, the Food and Drug Administration,

the National Science Foundation and the Executive Office of the President.[59]

Appendixed to the final EIS was an interim report of the Interagency Committee, which presented the results of it having examined the applicability of existing legislation to recombinant DNA research.[60] A number of problems had been encountered. For example, under the Occupational Health and Safety Act, the term 'employer' did not cover US states and their political subdivisions, unless they volunteered to adopt this status. Only 24 had done so. Self-employed persons were also excluded. Under the Toxic Substances Control Act, although recombinant DNA materials could be covered, the Act explicitly exempted the need to register small quantities used for the purposes of scientific experimentation or analysis. Indeed, similar problems arose with all of the existing legislation examined, such that the Interagency Committee concluded that although there was coverage within the legislation broad enough to include recombinant DNA, it 'would probably be subject to legal challenge'.[61]

From their survey, the Interagency Committee itself recommended elements to be incorporated in legislation. Both the production and use of recombinant DNA molecules, it felt, should be covered. They suggested that projects should be registered, facilities licensed, a single set of regulations should apply nationally which would pre-empt state legislation, and inspection and enforcement should be implemented. Only two abstentions (from the Council on Environmental Quality and the Justice Department) were recorded in a report which recommended legislation. As a result of the report, the Secretary of Health, Education and Welfare, Joseph Califano, had legislation developed. The Administration Bill was then reviewed by all members of the Interagency Committee, and by all departments and agencies. Senator Edward Kennedy, Chairman of the Subcommittee on Health and Scientific Research of the Senate Committee on Human Resources, and Representative Paul Rogers, Chairman of the Subcommittee on Health and the Environment of the Interstate and Foreign Commerce Committee, introduced the Bill in each House. Congressional hearings followed, and by the time of the final EIS, both House and Senate Bills were pending. Thus, of all the options outlined in the EIS, only Federal regulation was examined in depth. The actual failure of legislation to be completed is discussed in Chapter 6.

A number of letters received in response to the draft EIS were very critical of the limitations and superficiality of the discussion of alternative options. It was, for example, suggested that true evaluation required much more than a cursory summary.[62] Yet despite many comments, the final EIS did not adequately respond to them. A chapter given over to the comments on the draft EIS followed the same format, using the same headings as in the draft. Each section was then taken individually and a response was made to any criticism which fitted that particular category. Thus, when the five alternative options were returned to, there was no slot for those comments which were critical of the limited range of options, or to the superficial level of the discussion.

An important criticism raised by a number of respondents concerned the failure of the draft EIS to address the long-term evolutionary factors. The final EIS made this excuse:

The omission of this matter from the Draft EIS was based on several considerations, including the almost total lack of relevant scientific facts, the highly controversial nature of modern evolutionary theory, the consequent inability to impose a theoretical framework on the issue, and difficulties in analysing the arguments of those who have expressed serious concern with this matter...[63]

This is tantamount to arguing that because of uncertainty and extensive disagreement, the issue can be omitted from an assessment of environmental impact. Controversy should, on the contrary, necessitate its inclusion. A very brief reference to the issue was included in the final draft.

Criticisms applied to the guidelines rather than the draft EIS were simply forwarded to the RAC, which at this time was beginning the process of guideline revisions which along with legislative efforts is returned to in Chapter 6, where they are considered in relation to international activities. In sum, the legal requirement for a wide-ranging assessment of environmental impacts of somewhat controversial actions by the NIH was finally fulfilled. Much of the criticism of the draft EIS would suggest that something of a minimalist approach was adopted. Much of importance was omitted or glossed over.

BIOTECHNOLOGY IN THE UNITED STATES

Important as the discussion of safety was in the United States, it was not the only lead taken by that country in the field of biotechnology. By the mid-1980s it had become clear that the US had seen an enormous growth of industrial interest in applying the techniques of the new microbiology. Supported by favourable tax laws and the availability of venture capital, new US firms sprang into being to bask in the rays of the latest sunrise technology. These new companies were to include successful firms like Genentech, Genex and Cetus. Academics were seen to move into the industrial sector, even starting their own companies, while established industries sponsored both universities and the new start-ups, as they in turn rushed to acquire in-house expertise. New safety issues were also to ensue.

As the scale of use of genetic manipulation increases worldwide we can expect new problems of risk to arise. One immediate difficulty concerns the release of genetically manipulated organisms into the environment. It was not too long before the RAC faced its first requests to approve such proposals. Nor was it too long before the US again saw the regulatory issue come to prominence following pressure-group activity. In September 1983 it was reported that the RAC was being sued for approving environmental release experiments, without preparing an environmental risk assessment.[64] The challengers on this and many subsequent occasions were an individual, Jeremy Rifkin, and the pressure group he presided over, the Foundation on Economic Trends.[65] However, it was in May 1984 that Rifkin and other environmentalists achieved success in obtaining a court injunction halting a specific experiment proposed by Dr Steven Lindow, a researcher at the University of California, as environmental impact had not been rigorously investigated by the RAC. A mutant organism was to have been sprayed on the shoots of potatoes, which it was hoped could replace a natural variety, with the effect of reducing frost damage to crops. Following this decision and threatened to be the next Rifkin legal target, Stanford University subsequently postponed a proposed experiment involving the growth of genetically engineered corn plants.[66]

More broadly, Rifkin has attracted much support from many religious groups on the basis of an ethical view that genetic engineering represents unacceptable human interference in life processes, risks aside. Despite the emotions engendered by Rifkin's activities, both in favour and against his views, there is no doubt that he possesses some credence with respect to the environmental release of modified organisms.

The US Court of Appeals eventually overturned the lower court injunction against the Lindow experiment, while in the meantime the RAC prepared an environmental impact assessment before contemplating approval for further such experiments. Rifkin, in effect, was successful in having the RAC legally reminded of its duty to prepare assessments for all future proposed environmental releases. However, in the United States once again the question of NIH requirements covering industry surfaced. Private companies began to forward similar proposed experiments to the RAC. Although the Rifkin injunction did not apply to private companies who were not formally bound by the RAC guidelines, the Director, NIH, James Wyngaarden, decided to hold up 'approval' in any case, until environmental assessment was complete. One of the companies, Advanced Genetic Sciences, subsequently withdrew its proposal and was reported to be planning in future to deal with the Environmental Protection Agency (EPA), then in the process of extending its regulations to cover such field trials.[67]

However, not all applied uses of genetic manipulation require environmental release of the modified organisms involved. Pharmaceuticals and other industrial users of biotechnology might be more concerned with the large-scale production of biochemicals utilising modified organisms as host–vector systems. For this reason the RAC gave attention in 1979 and 1980 to the possibility of enhancing the guidelines to cover experiments or possible production processes which involved more than 10 litres of culture medium. After a series of meetings which finally agreed draft guidelines, the result was that the Director, NIH (still Fredrickson at that time) issued the proposals recommending that they serve as a model for those preparing submissions to the RAC. Three levels of physical containment were detailed, P1–LS, P2–LS and P3–LS. The P3–LS level not least required the establishment of a programme of health surveillance and the planning of emergency procedures to handle large losses of culture. No P4 equivalent was specified, although it was stated that recommendations for individual proposals requiring maximum containment would be established on an individual basis.[68] Actual RAC review of large-scale proposals from the private sector began in September 1979.

Safety in the United States and elsewhere thus remains an issue of importance as the applied dimension of genetic manipulation becomes more prominent. New hazards may need accounting for, such as the safety of commercial bioreactors, and difficulties of jurisdiction between various concerned government agencies also require attention. The sheer scale of commercial interest in applying genetic manipulation techniques and biotechnology in the United States makes what happens there of international interest. Indeed, many of the large US pharmaceuticals have themselves multinational interests. Any inherent safety issues yet to surface are as likely as not to come to light in the US, given the extensive range of activity there. Consequently the need for monitoring of industry's application of genetic manipulation may now overshadow the earlier focus on laboratory safety. With the uneasy distinction between industry and NIH-sponsored work, in respect of sanctions available to ensure guideline compliance, the extension of the jurisdiction of other agencies to cover recombinant DNA work may be the best option.

Compliance with the guidelines and large-scale recommendations cannot at present be guaranteed as the activities of industry are still only voluntarily tied to these provisions. The Congressional Office of Technology Assessment (OTA), however, has stressed the point that with the possibility of private legal action should any accident occur, industry would find any defence more difficult if it were shown that the industry had not adhered to the guidelines. In effect, the guidelines could represent the standard against which negligence might be assessed. In addition such negligence, if it occurred, could trigger subsequent legislation.[69] Problems with reliance on voluntary compliance were nevertheless identified by OTA. First, they identified the initial shortage of experience on the RAC in industrial techniques. Secondly, they indicated a problem also evident in the case of the UK. Industrial confidentiality could lead to the withholding of information by the firm concerned; or, thirdly, some RAC members opposed to secret sessions to assess confidential proposals would not attend, leading to a lessening of expertise. Fourthly, monitoring for guideline compliance after scale-up is limited. Finally, they noted the lack of penalties if continued compliance was not evident.

Aware of the growing complexity posed by the blossoming interest of industry in applying the techniques of recombinant DNA, the RAC moved towards the suggestion that the various agencies in the Interagency Committee be responsible for regulation. Generally speaking, the various agencies that have interests in any genetic manipulation work voluntarily apply the RAC guidelines to any sponsored research. However, the EPA in particular began to give formal attention to exerting regulatory authority over the commercial uses of genetically engineered organisms. The EPA had already acquired experience in biological pest controls, and combined with growing criticism of the RAC's lack of industrial experience, it moved towards developing experience in risk assessment in recombinant DNA. This move has not in itself been without criticism. The whole area of 'risk assessment' in technological activity, and specifically in biotechnology, is controversial for reasons outlined in Chapter 6. Further, the legal grounds for the EPA proposals were not definitive. Much centred on an interpretation of the US Toxic Substances Control Act. To apply the Act, and therefore justify EPA's intervention, 'new chemical substances' would have to accommodate organisms with altered DNA. This point is not at all clear. Should the Act be applicable, it requires manufacturers to give EPA 90 days' notice of intended production of a new substance and provide all relevant safety details. EPA could then if it wished require further safety assessment. Despite the potential promise of bringing industry more formally under some regulatory controls, the Act would not cover research and development use of new substances. A comprehensive legal package covering all recombinant DNA work is not, therefore, implied.[70]

SUMMARY

At this stage a limited response to the substantial questions raised in the Introduction is possible. In the first instance there is some evidence that political biases and constraints existed in terms of US institutional procedures. In terms of designing and implementing the guidelines the views of scientists and administrators specialising in the promotion of science were dominant. The NIH was working within a notable constraint of not impeding the development of recombinant DNA work more than necessary. Many debates did, however, occur involving a wider set of participants, at various decision-making levels within the US, but it is my conclusion that although the many viewpoints were acknowledged by the NIH at the Federal level, their influence was limited to being reflected in the general level of caution. Wider influence on the actual implementation of guidelines was thus only really evident in the more localised debates such as at Cambridge, Massachusetts. In the longer term the dominance of the 'experts' would face further challenge in the threat of legislation and with the widening of participation on the RAC.

Undoubtedly, control options were not analysed exhaustively by the US institutions. The discussion of the Environmental Impact Statement clearly illustrates this point, suggesting that the limited options examined were somewhat superficially addressed. Indeed, the NIH simply followed a course originally asked of it by the Berg letter itself. Guidelines simply to enable the work to continue were the order of the day all along. In part, the initial narrowing of the whole issue area to questions of safety explains the concentration on certain types of option, assuming the continuance of the research. Had official US bodies made early attempts to consider positively some of the more ethical questions about the impact of the ability to manipulate the genetic basis of life itself, then events might have been quite different, especially if the consequence was even more public scrutiny.

It was not until the formation of the Interagency Committee some two years and three months after the Berg letter that any serious thought was given to going beyond the adaptation of existing frameworks in the implementation of controls. The guidelines did bring innovation in the context of regulating a new area of research, but the procedures used required minimal change to the body which promulgated them, the NIH. This need not be a criticism if the body which had effective authority was most suited anyway. It is argued here that it was assumed that the NIH was suited by the scientists involved and this was supplemented by a perceived fear, from the scientist viewpoint, of non-scientist dominance if the NIH was not used. As will be shown, no major changes to the implementation procedures for controls were to be forthcoming as scientist pressure-group activity and changing circumstances were to ward off legislative developments. Even the belated first real move towards extending procedures, the efforts by the Environmental Protection Agency to cover commercial applications of genetic manipulation, was itself founded on an attempt to adapt existing regulatory provisions.

Although a number of non-scientists made comparisons between genetic manipulation and nuclear technology, no institutional efforts were made in the US to make systematic applications of knowledge from other fields regarding technological safeguards in general.[71] This is a point that should be taken further by way of comparison with other countries, as arguably there might have been much to learn by making reference to the experiences of other technologies such as the chemical and nuclear fields. These comparisons need not have been restricted to the logic and techniques of assessing and containing hazards, but could also have focused on questions of accountability and legitimacy in the making of decisions and policy, to the extent perhaps of improving on earlier experiences.

Communications patterns are probably best addressed at the transnational level. Nevertheless, some points are of note concerning the US. The central bodies of the RAC, ORDA and the Office of the Director, NIH, were very well linked on technical and implementation criteria. Overall, the non-scientists had great access to information through the open publication of such material and the public operations of groups like the RAC. Press activity was substantial in comparison with European states, and interest groups made much use of such sources. Input into the decision process was also substantial, although biases in the handling of it were evident.

These comments are indicative of the overall issues to be returned to in a more international assessment, and some points will be raised concerning the US in making comparisons with the British experience, and later in considering the pattern by which guidelines were revised internationally.

4 Response to Concern: The UK Model

Some of the reasons given for examining institutional responses to the expressed concerns about recombinant DNA apply to both the United States and the United Kingdom. Both states witnessed urgent investigation of how to proceed safely with the research, and both states found their guidelines, and procedures for their implementation, borrowed or adapted by others. It is important, therefore, to consider the developments within the UK which paralleled first responses in the US. Although both the UK and US approaches influenced other states, they were in fact quite different from each other, reflecting differences in policy and national requirements. In particular, there were major differences in the way in which the overall control procedures were implemented, especially regarding the functions of central committees. The RAC was constituted and operated quite differently from the UK Genetic Manipulation Advisory Group (GMAG), so much so that the acronym GMAG ('Gee-Mag') became an international byword to describe a particular approach to the problem, although more recently its name and locus of responsibility have changed. A degree of comparison between the two approaches is therefore necessary.

Because of the speed with which both the US and the UK responded to the Singer–Söll and Berg letters, they established lead positions, with many states preferring to await the results of their assessments. If uncertainty has been identified as a key variable then it was at its height when the first UK working party examined the issue. Even states that devised their own guidelines were on the whole more hesitant. However, the lead of the US and UK was not only confined to assessing the issues and producing operational control measures. These two states significantly led in terms of the actual utilisation of the recombinant DNA methods. By 1978, some 50 US and 45 UK laboratories were exploiting the new techniques.[1] As with the US, for analytical purposes the UK operationalisation of control is seen to operate within a distinct system, although it is acknowledged that activities within other states, notably the US, were of some influence. Again, an historical overview will provide a description of the development of guidelines and their implementation, deferring much of the analysis of consequences to later chapters.

In many ways, the task of describing how the UK as a whole responded is more straightforward than in the case of the US. Being a somewhat smaller country facilitated the development of a more centralised method of assessing the issues and implementing the control system. Regional difficulties did not arise like those resulting from the federal structure of the US. Indeed, it will be shown that the ease of central administration of control in the UK led to a particular mode of operation which in some ways was also seen as suited to other European states. Also, and in part attributable to the different internal setting of the UK, there was less public and popular press discussion of the issues. The more open form of government in the US, coupled with traditions of interest groups lobbying Congress and decision-makers in general, fostered an environment more likely to promote wider public discussion. Added impetus to this derived from the fact that it was mainly US scientists who had taken first actions in publicising the whole issue area.[2]

Overall, documentation on issues as they developed in the UK was considerably less than in the US. With a less obvious debate involved, there was correspondingly less need for quite the same volume of official material to be published. Communication was more obviously informal as the community of scientists and relevant administrators was less dispersed. However, the few official reports which were issued must be seen as of great importance, both within the UK and within the wider international context. In effect, the participation leading to the development of a UK approach to control of the recombinant DNA techniques was narrower than in the US. This point must nevertheless be separated from the fact that, in *implementing* the resulting guidelines, participation was *wider* than in the US, both in terms of the number of agencies involved in policy and in terms of the interests represented on GMAG. Yet in the development of institutional responses the activities of UK working parties were critical steps. Publication of the Berg letter in *Nature* on 19 July 1974, one week before publication in the US, drew the rapid announcement on 26 July that a working party would examine the issues. It was to be chaired by Lord Ashby, the Master of Clare College, Cambridge, under the auspices of the Advisory Board for the Research Councils (ABRC).

Part of the reason for the speed of response can be explained in the light of the consequences of a smallpox outbreak in London over a year beforehand. The outbreak was important in creating an awareness, or at least providing a recent reminder at that time, of the need to contain hazardous viruses. It also created an institutional response in its own right, that was in many ways to interact with subsequent responses to genetic manipulation. The details of the smallpox case are well documented as a consequence of a report published by a committee of inquiry.[3] Of note is that the origin of the outbreak was a failure in the containment procedures of a research laboratory, at the London School of Hygiene and Tropical Medicine. A technician from a different laboratory had, on 28 February 1973, witnessed an experiment involving the harvesting of smallpox virus. On 11 March the technician fell ill and she was transferred to hospital by her doctor a few days later with suspected meningitis. Yet it was not until 23 March that smallpox was diagnosed, and then only after a combination of factors. Relevant information had been spread between a number of people, who were each in ignorance of the other. Further, the technician had been placed in a public ward prior to the correct diagnosis, and as a consequence two visitors to a neighbouring patient contracted the virus and subsequently died, although the technician recovered.[4]

A significant point was illustrated with the London smallpox outbreak, and a second outbreak in Birmingham some years later. Once a dangerous organism escapes from its confinement, the consequences might not immediately be apparent. If difficulties could occur with known pathogens, then this would beg the question of monitoring the impact of unknown pathogens, particularly if, in recombinant DNA work, there was a delay in the 'expression' of the cloned DNA in its host.[5] In the case of the 1973 outbreak, the smallpox virus was only recognised as a result of worry on the part of the technician's superior, who took a skin sample from her during hospital visiting hours! The identification of the two visitors who subsequently died as being infected was a result of some inspired deduction by a social worker who read about smallpox in the press.

At the time of publication of the Berg letter, the lessons of the smallpox case were being processed. As a result of the inquiry into the case, the Secretary of State for the Social Services set up a working party, under Sir George Godber, with the following brief:

To consider whether there are organisms capable of causing communicable diseases that require measures to be taken in laboratories or elsewhere additional to those now recommended in order to prevent infection in man or in animals and to make recommendations as to the measures required.[6]

The UK was, therefore, institutionally discussing biological hazards when the specific issue of genetic manipulation emerged. Thus the Ashby working party moved quickly, in part as a result of this earlier experience and the existence of the working party examining dangerous pathogens. However, the rapid response was also due to the transnational discussions which occurred between members of the Berg letter group and others over the months prior to the publication of the letter. The Medical Research Council (MRC) in particular began to consider an official response before the letter, and in July 1974 sent confidential letters to its laboratory directors effectively banning all the types of experiments questioned by the Berg group.[7] Thus a British ban was very quickly introduced pending the investigation of the problem by the Ashby working party.

COMMITTEES OF INQUIRY AND A CODE OF PRACTICE

The Ashby Report

Two preliminary points should be noted regarding the Ashby Report. First, the investigation of the issues was over a very short time period, at the request of the ABRC which hoped for an opinion before the autumn. It was left for the Ashby working party to decide whether or not it would be a final or interim report.[8] Produced after only five months, the Report became designated by Ashby as a 'consultative document', rather than in any way as a final report. He hoped that it would 'stimulate discussion both in the scientific community and by the general public'.[9] Secondly, the Report was published before the Asilomar II meeting to enable something of the UK position to be determined and fed into this very important international meeting. Any criticism of the report must therefore take account of its expressed intention of fostering domestic and international discussion. A later report, examined below, would work on UK policy. The Ashby working party operated within the following terms of reference:

To assess the potential benefits and potential hazards of techniques which allow the experimental manipulation of the genetic composition of micro-organisms; and to report to the Advisory Board for the Research Councils.[10]

It would appear that the remit included a rudimentary idea of risk-benefit assessment, rudimentary in that no clarification of the criteria by which risks and benefits could be compared and assessed was given in any rigorous fashion. In the context of a technology displaying features of low-probability, high-consequence risk, it is argued here that risk-benefit assessment would be inescapably politicised, in reflection of the different values and perceptions involved. The Ashby Report, although it acknowledged the wider non-scientific debate already in evidence, did not take up any of the issues explicitly. Essentially, what the Ashby Report did was, in the first instance, to present a summary of the techniques to date and their conjectured potential benefits.[11] Hazards and potential methods of their reduction were then outlined, from which conclusions and recommendations were drawn.

Taking the Report as a whole, the overall feel is one of subjectivity. In effect, hazards were listed, benefits were listed, and, based on evidence from scientists in related fields,[12] a subjective balancing exercise was conducted. The working party to its credit acknowledged its dependence on 'experts', but failed to justify its confidence in the type of expert called to give evidence. At the time only a handful of scientists worldwide were engaged in recombinant DNA work, although many anticipated using the techniques. The latter group more accurately describes the witnesses called. In general, the Report did not sufficiently stress the conjectured nature of both the risks and the benefits. Conjecture itself implies uncertainty, yet the report emphasised the 'informed' status of the witnesses.[13]

Many would probably argue that the fact that a working party of this sort was set up at all was a stimulus for discussion.[14] Nonetheless, this does not belie the importance of accuracy in such a report. Assumptions based on inaccuracy were apparent in at least one example. It was stated that 'our general philosophy for defences against potential hazards is that they should not be employed when it is patently unreasonable to do so, as, for instance, in most experiments on plants'.[15] This statement displayed ignorance of the need for safeguards at the plant level, for example to avoid potential crop infections. Indeed, a report following on from Ashby argued that 'suitable measures of containment for ... plants will be needed'.[16] As the second report was directly important in the establishment of UK containment procedures, plant experiments were incorporated in the UK guidelines, as with those of the US.

In addition to accuracy, a discussion document can have influence through what it declines to discuss. Besides avoiding ethical, political and social factors, often omitted in officially sponsored reports, the Ashby Report intriguingly raised a 'hazard' and then declined to discuss it:

We mention one other hazard, although only in passing because it is not within our remit. The question may be asked whether the techniques we are assessing could be used in bacteriological warfare. We have no special knowledge of this field; but we can conjecture possible malicious uses for these techniques.[17]

It is curious that the Report claimed that this hazard was not within their remit, when the terms of reference above merely suggested an examination of potential 'benefits' and 'hazards'. Definitions of benefits and hazards were their responsibility, and if they chose to term usage of the techniques for bacteriological warfare purposes as a hazard, then they would have every right, if not duty, to consider this. As with the US, the possibilities of recombinant DNA techniques for weapons development were never discussed in great detail within institutions charged with developing safeguards.[18]

Perhaps the dominant recommendation of the Ashby Report was that the moratorium on using recombinant DNA techniques should be 'no more than a pause' and that work should continue. It should, however, be noted that, as in the US, initial response in assessing the issues raised by the scientists who voiced their concern was taken under the auspices of bodies concerned with the promotion of basic research, and not specifically with the control of risk. In the US this had been the National Institutes of Health, while in the UK the Medical and Agricultural Research Councils had called upon the ABRC to investigate, and hence the Ashby working party resulted. Subsequently, the Department of Education and Science would take up the issues, again a body charged with promoting science.

Given that the Ashby Report was an initial and speedily produced consultative document, with some inherent weaknesses, it is nevertheless important briefly to assess its impact. Many of its recommendations were to be adopted, for example, the establishment of a 'central advisory service' and the use of biological safety officers in research laboratories. These were made operational after the next working party took implementation considerations further. One science journal carried a leading comment summing up its reaction with the simple heading; 'Not Good Enough'.[19] The nub of its criticism was that work should continue with 'not even a voluntary pause while ... safeguards are developed'. In particular, the author, Bernard Dixon, saw this as disquieting on the eve of an international conference expressly called to consider the whole question. However, Ashby did propose that 'rigorous safeguards' be applied, and as it happened Asilomar II was to recommend some interim measures regarding temporary guidelines. What was worrying was that the Ashby Report saw the use simply of containment practices which were applicable to any pathogen, in conjunction with good laboratory safety practice, as adequate. Asilomar II was more cautious and specific. Ashby talked of degrees of hazard applying to pathogens in general, but made no assessment of how to rank degrees of risk in relation to types of recombinant DNA experiments. Instead, the Report implied that each individual biological safety officer should give guidance on the precautions necessary. The qualifications of such a person were not discussed, although as mentioned a central advisory service was recommended, and, in addition, it was suggested that somebody draw up a code of practice. An editorial in *Nature* summarised the problem thus:

The real question is whether it is possible to impose from scratch, on scientists and technicians who have not been used to them, the disciplines of institutions that deal on a day-to-day basis with pathogenic organisms...'.[20]

It was not until the Williams Report that such issues were rigorously considered in the UK, and this was published 19 months later. The Ashby Report was at least important in engendering discussion on the issues, for example within the science journals and, in conjunction with the Asilomar II recommendations, it prompted the subsequent Williams working party to be set up. However, one final point regarding the Ashby Report deserves mention. Commenting on the then recent London smallpox outbreak, the Ashby Report said: 'The dramatic response to any failure of containment illustrates how rare such failures are.'[21] This was a common argument of defenders of nuclear power in the aftermath of Three Mile Island – an argument which must surely come under question following the further lessons of Chernobyl. Apart from underestimating the luck that led to the identification of the virus itself as causing the technician's illness and the identification of visitors to the ward, infected later, the comment fails to appreciate the particular problem of perception involved in all such issues of low-probability, high-consequence risk. In general, the report seemed to rely too much on the idea of good but routine precautions, underestimating the uncertainties involved.

Thus a somewhat hastily produced document was the overall result, and these criticisms reflect that. Yet the report was of influence on both sides of the Atlantic in stimulating the impression that the moratorium should end quickly. One analyst, Edward Yoxen, has described the Ashby Report as having failed to encourage discussion on wider policy issues at 'a pivotal moment in the emergent debate'.[22] Not least this failure had a transnational dimension. The Ashby Report influenced Asilomar II, organised as a non-governmental conference in another state, which in turn influenced UK policy.

At this point, the scene was set in the UK for the second stage in the overall institutional response. This was to include, among other things, the development of guidelines, implementation procedures, and an increase in complexity at the institutional level as more government departments became involved. In turn, this was to cause interdepartmental rivalry, but more importantly raise questions of efficiency and legitimacy. Until the UK guidelines became operational, safeguard control had rested with the MRC, but, on 6 August 1975, the Department of Education and Science (DES) announced through a press notice the formation of a second working party.[23] Stating the conclusions of the Ashby Report, the Secretary of State, Mr Fred Mulley, went on to give his reasons for establishing the working party. Acknowledging that scientific bodies had endorsed the Ashby Report, and that the Godber working party on dangerous pathogens had also considered it, Mr Mulley outlined the government view. The government accepted that it had a responsibility to ensure the availability of authoritative advice and guidelines to enable work to continue in appropriate places and with stringent precautions. Mr Mulley also noted:

At the same time we believe that the potential hazards associated with certain types of experiment are such that it would be appropriate further to examine the possibility of applying to them controls of the kind recommended in the Report of the Working Party on the Laboratory Use of Dangerous Pathogens.[24]

Until advice from the new working party was available, Mr Mulley asked that the various research councils and others concerned did not proceed with work 'already identified as involving potentially serious hazard'. Thus, as in the US, there was a relatively early decision taken in official circles to work towards the development of controls for genetic manipulation research.

The Williams Report

The working party was to be under the chairmanship of Professor Sir Robert Williams, Director of the Public Health Laboratory Service, London. Williams was one of four members who had sat on the Godber working party which had examined dangerous pathogens and had himself also served on the Ashby group. There was, therefore, some significant linkage between the three working parties. Taking note of both the Ashby and Godber Reports, the Williams group was instructed:

(a)   To draft a central code of practice and to make recommendations for the establishment of a central advisory service for laboratories using the techniques available for such genetic manipulation, and for the provision of necessary training facilities;

(b)   to consider the practical aspects of applying in appropriate cases the controls advocated by the Working Party on the Laboratory Use of Dangerous Pathogens.[25]

Indeed the announcement in August 1975 of the new working party came fairly close on the heels of a meeting of some 100 scientists, held in Oxford, over the weekend of 12 July 1975, at which, it appeared, patience regarding the moratorium was beginning to weaken. Several participants had indicated the intellectual pressure which was building up to get moving again in the field a year after the Berg letter. Although the press had been excluded, *Nature* carried an editorial which both criticised this fact and suggested that at the meeting it was conceded that some scientists had already begun work.[26] It appeared that some form of code of practice was urgently required.

It is interesting that the Report which was finally produced emphasised the differences between handling known dangerous pathogens and the many types of recombinant DNA experiments. The former involved the application of well-known precautions against a small number of easily identifiable and well-characterised agents, while the latter would involve each experiment being assessed independently. Ironically, some years later pressure developed to adapt the new genetic manipulation procedures of control to dangerous pathogens as a consequence of a further smallpox outbreak.[27] In reaching its conclusions, the Williams working party had invited evidence from a much wider selection of interested parties than had Ashby. Of particular significance, it consulted representatives of trade unions, the Confederation of British Industry (CBI), government departments, the Committee of Directors of Polytechnics, as well as those with scientific interest or who intended to use or develop the techniques. Such wider consultation gave the report a sounder and more legitimate base. The main results were the production of a *code of practice*, or guidelines, and the designation of a central advisory body. These can be examined in turn.

Important as the UK code of practice was, it did not differ greatly from the US guidelines. That is, they shared similar conceptions of physical containment categories, covering four in number. They were not, however, identical and a report of the meeting of the EMBO Standing Advisory Committee on Recombinant DNA includes a comparison.[28] In the UK guidelines no equivalent was specified for P1 in the US, Category I (UK) was more stringent than P2, Category II effectively equalled P3, Category III had no US equivalent, while Category IV was equivalent to P4. Differences were sufficient for other states adopting either to have some choice. Both sets of guidelines also laid emphasis on the role of biological containment, but the Williams Report reserved far more emphasis for the role of physical containment. Regarding the possibility of enabling increased biological

containment to offset physical containment requirements, the
Williams approach was to argue:

We assume that there are conditions of biological containment and nucleic
acid purity that will allow an experiment to be moved from one category to
another but these cannot be absolutely defined without reference to the
individual experiment.[29]

The report did not specifically categorise biological
containment at all and, although physical containment was
categorised, there was no comprehensive allocation of
experiment types to the containment levels involved. Those
few experiments which were collated with the four containment categories were not to be seen as anything more than a
guide. The UK approach, it was intended, would utilise a
central advisory service which, upon notification of proposed
experiments, would allocate the containment category. It was
hoped that in this fashion a body of 'case law' could be built
up offering greater flexibility.[30]

Overall, these proposed UK guidelines, which were
eventually implemented, were tougher than those of the US, in
terms of physical containment requirements, but unlike the US
guidelines did not completely ban any experiments. However,
an important shared element of the Williams Report and the
NIH guidelines document was the emphasis on the importance
of appropriate training for workers who would use
recombinant DNA techniques. In particular, this recommendation was in recognition of the fact that many researchers were
likely to move into the field perhaps unaware of the routine
techniques of medical microbiology.[31] It was argued that
training would be one of the responsibilities of a Biological
Safety Officer, one to be appointed to each laboratory
concerned. His responsibility in general would be for pre-
cautionary measures and he in turn would require training. A
number of training courses were subsequently run. Thus, the
code of practice was to include an important local element in
the implementation of safeguard controls. It was not as
extensive, however, as with US local Institutional Biohazard
Committees.

In essence, the main difference between the first US and UK
guidelines was that the former were designed in such a way

that the leading researcher in any laboratory could consult them personally and determine the appropriate containment for his experiment. The US guidelines thus included a comprehensive listing of experiment typologies. The more flexible UK system enabled the central advisory committee to allocate containment on the basis of: the nature of the experiment; the laboratory's facilities; the experience, ability and training of the research workers, technicians and the Biological Safety Officer. Records of each laboratory would be kept, and annual reports were proposed. With the built-in flexibility of the system, revisions of containment recommendations could be readily made in the light of experience and changing perceptions. In the US, such developments would necessitate guideline revision.

Central screening of all experiments, although occasionally suggested,[32] was not seen by the NIH as appropriate for the US. A problem of scale existed, both in terms of the likely future number of laboratories involved and the sheer size of the US. Local pressure was also evident in some US states and cities regarding the right to legislate for controls more rigorous than those advocated by the NIH and suggested in Federal legislative proposals. The dominant perception appeared to be that central vetting of all experiment proposals would be too cumbersome. This was not the case in the UK and an appropriate central committee resulted.

THE GENETIC MANIPULATION ADVISORY GROUP

Recommended in both the Ashby and Williams Reports, a UK central advisory committee was established under the title of the Genetic Manipulation Advisory Group (GMAG). Eventually to be an important body in the UK control system, it was also a notable development in the whole concept of control of safety in scientific research. Its acronym, as mentioned earlier, was to become an international byword.[33] Although GMAG was outlined in the Williams Report as part of the overall package, the report itself faced a considerable delay in publication. The explanation for that delay is best made after GMAG and its relationship with another body, the

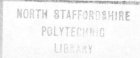

Health and Safety Commission (HSC) is described. Essentially, the delay occurred to allow the HSC to produce a further document defining 'genetic manipulation'.

It was planned from the beginning that GMAG, in order to 'command the respect of the public as well as of the scientific community, including scientists in industry', should comprise a wide range of representation.[34] Involved scientists, industry, employees and the public interest were all to be represented. It can be argued that GMAG was a very innovative proposal, in that scientific work at the research level would be monitored, in respect of a code of safety practice, by a group in which scientists directly interested in using the techniques were to be in a minority.[35] Other states, including the US, were similarly to raise the question of broad participation on their advisory committees. However, some notable contrasts can be made with the advisory group established at the behest of the Godber Report on dangerous pathogens. By the time the Williams Report was published, GMAG's cousin, the Dangerous Pathogens Advisory Group (DPAG) was already in operation. Although the Williams working party considered that DPAG's functions might be extended to cover genetic manipulation, this was rejected on the basis of the different procedures involved between known pathogens and the uncertainties involved in recombinant DNA. DPAG when established as an advisory group saw these differences reflected in its structure. Whereas GMAG had broad participation, DPAG was more specifically based on experts in dangerous pathogens.

In composition, GMAG was to have 19 members: four trade unionists representing employees working in laboratories carrying out genetic manipulation; four members representing the 'public interest'; two representing management; eight representing scientific and medical experts; and the first chairman, Sir Gordon Wolstenholme.[36] Of these, some controversy was to surround the appointment of the representatives of the public and employees. Public interest representatives were to be invited to join GMAG by the DES, but without any published criteria underlying their selection. The difficulty was highlighted at the end of GMAG's first two years when some replacements in the membership occurred.

One of the public interest representatives, J.R. Ravetz, a Reader in the History and Philosophy of Science at Leeds University, did not want to leave GMAG, but was requested in writing to stand down. Ravetz was particularly active and not entirely uncritical of control policy in general.[37] The reason given was that new appointments were necessary to 'ensure the balance and continuity' of the group. Yet two of the other public interest representatives were in fact keen to leave, which would leave only one of the original four. There would seem to be no reason why Ravetz could not have stayed on. If 'continuity' were to be stressed, then it should be said that it took some time to gain experience and understanding of the more technical and scientific issues involved in GMAG's activity. Three replacements would represent a significant discontinuity.

Trade union representation from the beginning involved some active lobbying by at least one union, the Association of Scientific, Technical and Managerial Staffs (ASTMS). ASTMS had begun to show an interest in recombinant DNA issues prior to GMAG being set up, as a result of one of its members, Professor Robert Williamson, a biochemist, having relevant contacts in the US. For Donna Haber, who took considerable interest in genetic manipulation on behalf of ASTMS, the recombinant DNA case provided an opportunity for preventive action, rather than court action after the event, with which ASTMS was more familiar. Thus ASTMS lobbied the Williams party, both to establish a central advisory committee and to include trade union representation on it.

In practice, representatives of both trade unions and industry were to be recommended to the DES by the Trade Union Congress (TUC) and the industries concerned. At the time GMAG was set up in late 1976, the Secretary of State for Education and Science was Shirley Williams, who in retrospect has stated that both she and her department consulted widely on all appointments. She did, however, qualify this: 'they were all consulted. That is not the same thing as accepting recommendations made by officers [of the various organisations involved] without further question.'[38]

Apart from Ravetz, others saw reason to question appointment policy. In particular, the Association of

University Teachers (AUT) wanted a member on GMAG, but were unsuccessful in obtaining TUC backing.[39] A rather forceful letter to the Clerk of the Sub-Committee of the Select Committee on Science and Technology, which in 1978/79 examined recombinant DNA research, went further on the question of participation. Professor S.J. Pirt of Queen Elizabeth College, the University of London, suggested the inclusion of representatives of the main scientific organisations in the field, including the Genetics Society, the Society for General Microbiology, the Biochemical Society and the Institute of Biology.[40] Pirt was somewhat critical of appointments being made by civil servants 'trying to have their own way as usual'.

GMAG, because of its innovative nature, had no real precedents from which to work in terms of the selection of representatives. Shirley Williams had been conscious of the difficulties in obtaining a suitable balance, and had been aware of the dangers of appointing too many scientists involved in the work, because being few in number they would effectively have regulated themselves.[41] Despite these problems, GMAG was a brave attempt to form an advisory group of some legitimacy. It was also quite successful.

In contrast, the underlying philosophy for the composition of DPAG was that it should be 'a small independent body of experts consisting of individuals whose experience would command the confidence of those working in laboratories.'[42] It was not until December 1978 that an offer was made to allow one trade union representative to sit on the group.[43] Yet, despite the differences with GMAG and the fact that it was administered by the Department of Health and Social Security (DHSS) rather than the DES, many of the issues which they faced were similar. For example, both groups were concerned with the safety of employees, the operational monitoring and implementation of a code of practice and the maintenance of public confidence. Confidence in DPAG was, however, to be shattered in a dramatic fashion in August 1978, with the result that calls arose for it to be reformulated along the lines of GMAG.

The event of such significance was another smallpox outbreak, this time at Birmingham University, and after the laboratory involved had been vetted by DPAG. DPAG had sent an inspector to Professor Bedson's laboratory in February 1976, at a time when no work involving smallpox virus was under way. The subsequent report into the outbreak criticised the efficiency of that inspection in not examining adequately the details of the laboratory and the proposed techniques to be used in future experiments.[44] Details of this confidential report, however, first came to light only after ASTMS published their copy of it, which was then given further publicity in the science press.[45] One of the problems was that the inspector had decided to overlook a number of deficiencies in Bedson's laboratory, such as a lack of an airlock, a shower, a double autoclave and changing facilities, on the basis of accepting Bedson's reputation as an experienced and safety-conscious virologist.[46] A significant factor had been, therefore, the inspector's 'working colleague' relationship with the head of the laboratory. Indeed, most of the visit was spent discussing smallpox work in general.

Other factors were also involved in the smallpox outbreak, including the approach taken by the World Health Organisation (WHO) in corresponding with Bedson. *Without the knowledge of the university* the WHO had informed Bedson that it intended to cease supporting his work after the end of 1978. WHO recognition was vital for the continuation of the research, and the effect of the deadline was to make Bedson speed up his work. In addition the WHO sent three inspectors who were critical of the safety procedures, but in the event the WHO allowed his promising work to continue, subject to the laboratory ceasing such work when the deadline expired. The tragedy of the whole situation was the subsequent infection and death of a photographer, who worked at the university, and the suicide of Bedson.

Calls for changes in the composition of DPAG soon appeared.[47] DPAG was eventually reformed with the addition of trade union representation. For a time, therefore, the UK had two contrasting examples of central advisory committees, one dealing with known biological hazards and one with new conjectured biological hazards of unknown risk. The latter, of more concern here, to an extent, became a pattern for a reformed DPAG, a group that some, including the Godber working party, had originally thought could have extended its sphere of operations to include genetic manipulation.

Despite the importance of GMAG in implementing the UK code of practice, it was not the only organisation involved. It has already been said that there was some delay in publishing the Williams Report, due to the Health and Safety Commission, an organisation central to the application of the code of practice.

APPLYING THE CODE OF PRACTICE

Provisions of the Health and Safety at Work Act
The Health and Safety Commission was established on 31 July 1974 under the Health and Safety at Work Act (HSW Act). It was to consist of between six and nine members from the TUC, and the CBI and local authority organisations appointed by the Secretary of State for Employment.[48] Its function was to monitor safety and health provisions in general in the context of 'work'. In addition, it was to have an executive arm charged with implementing its policy and acting as an inspectorate, the Health and Safety Executive (HSE).

Although the Ashby Report had overlooked the provisions of the HSW Act, this was corrected by the Williams Report. Williams explicitly suggested that regulations be made under the Act to require laboratories to submit experimental proto-cols to GMAG.[49] Publication of the Williams Report itself was delayed three months to give the HSC an opportunity to produce a consultative document covering draft regulations.[50] Both were published on the same day, but were not equally received by the scientific community. The Williams Report was generally accepted, while the HSC consultative document was heavily criticised.[51] Two major points of criticism stood out. The first concerned the attempt by the HSC to define genetic manipulation as follows:

No person shall carry out any activity intended to alter, or likely to alter, the genetic constitution of any organism unless he has given to the Health and Safety Executive notice ... of his intention to carry on that activity.[52]

The problem was that the definition incorporated the

traditional tools of genetics used for many years and could even be seen to cover activities such as spraying roses with pesticides, making yoghurt and even human procreation!

Secondly, the HSC document appeared to many to reduce the importance of GMAG as recommended by the Williams Report, by breaking the 'close link' envisaged between laboratories and GMAG. The HSC argued that laboratories should statutorily report to the HSE, and that GMAG should become simply an advisory body to the HSE, although it would still assess proposals given the lack of HSE expertise. Part of the problem was emerging political manoeuvres between the different overseeing departments involved, namely the DES (GMAG), the Department of Employment (HSC) and the DHSS (DPAG).

With its 'consultative document' the HSC had actually invited critical response.[53] This they got in abundance, although to be fair there was a degree of overreaction on the part of scientists and scientific organisations. They feared, in particular, a legal bureaucracy developing. Whatever the merits of the HSC definition, it had become clear that defining genetic manipulation with precision was a difficult task. Even ASTMS subsequently tried and failed,[54] and it was not until GMAG was constituted that the HSE and HSC got their definition after seeking its advice.[55] Final regulations were not in fact introduced until 1 August 1978, some 20 months after GMAG first met.[56] With this delay, the system in the meantime operated on a voluntary basis and by mid-February 1978, GMAG had received 102 proposals from 27 centres, including industrial laboratories. Four proposals were for Category IV containment, 27 for Category III, 40 for Category II and 31 for the lowest category.[57]

Once the HSC regulations were in force, however, notification had to be made by law to both the HSE and GMAG, and the definition of 'work' under the HSW Act was extended to cover genetic manipulation by any person, self-employed or non-employed (for example, a research student). This was an important difference between the UK and the US. The UK had a sufficiently flexible existing legal framework within which the Williams code of practice could be supported. Of further importance was that the legal provisions also applied to industry. From the beginning, the UK approach was intended to encompass all users of recombinant DNA techniques. Thus, the UK avoided the difficulties faced in the US regarding the limitation of the NIH guidelines to research wholly or in part sponsored by the NIH, or other government agencies with their agreement.

Because of the overall differences between the UK and US systems for operationalising control of safety, it is more appropriate to take each approach as a 'package', a conclusion that GMAG itself arrived at.[58] Implementation of the guidelines was much more centralised in the UK system, and this was reflected in the operations and the structure of GMAG. However, particular responsibilities accrued to the RAC in its role of developing and revising the guideline conditions for recommendation to the NIH. This function was more comparable with the endeavours of the Williams working party which initially devised the UK code of practice. Of note, though, in terms of policy-making, it can be seen that in the US the main souce of policy was the NIH. In the UK, by contrast, there was an additional problem of competing departmental roles making the focus of policy-making less clear.

The joint functions of GMAG and the HSE were at the heart of the UK package, although their overseeing departments differed. GMAG provided advice to both the laboratories and the HSE regarding containment, but it was the HSE which was to have the responsibility for inspection and, if necessary, legal enforcement. The function of inspection was a compulsory requirement before any Category III or IV work could be carried out. By January 1979, the HSE had three specialist inspectors in the field of genetic manipulation who in turn could call on the support of the HSE inspectorate's wider team, already functioning under the HSW Act in examining laboratories in general. In addition, the HSE could bring in the resources of the Employment Medical Advisory Service and the Factories' Inspectorate to deal with enforcement matters arising out of inspections. In all, the HSE felt it had access to sufficient expertise.[59]

It was, however, recognised that some problems might arise if an experimenter, cleared to carry out work at one containment category, upgraded the experiment without notifying GMAG or the HSE. Categories I and II did not require prior inspection, and the possibility therefore existed for a Category II experiment to be revised such that it should really be carried out under Category III. In response to such a possibility, the HSE took the line that not to notify would involve an unlikely level of irresponsibility on the part of the experimenter and in any case he would run the risks of the HSE's unannounced visits which all laboratories could receive. Supporting this was the general need in science to publish results for peer review, which itself could bring attention to violations. Nevertheless, resources at the disposal of the HSE were limited and there was no formal provision for regularised follow-up visits to laboratories. Against this, however, was the fact that initial visits were comprehensive to the extent of thorough interviews with local safety committees and employees' representatives, who would all be encouraged to maintain vigilance on laboratory work. It was thought that in the light of difficulties in prosecuting Birmingham University over the smallpox incident, legal sanctions might be difficult to apply.[60] Whatever the outcome of legal cases of this sort, the institution involved would invariably face much press publicity as in the case of Birmingham – itself a deterrent. Not least, a committee of inquiry might ensue. In all, the UK had available a much tighter system of ensuring compliance than in the US with its local level of enforcement supported by threats to funding in some cases.

In sum, the prior existence of legislation under which regulations could be drawn up was a great advantage, enabling a cautious approach of simply introducing requirements for notification. By redefining 'work' to cover the relevant research activity, the full force of the HSW Act could be applied. Yet in all this there were no regulations stating that advice had to be followed. Indeed, GMAG had been recommended to issue a disclaimer when giving advice, for fear that legal liability could go beyond them to the Secretary of State for Education and Science. The group itself, it should be said, had been willing to stand by its advice in law.[61] This was, however, a greater constraint in the US where the NIH did not give advice on individual experiments, in part because of the possibility of legal action against them if anything went wrong. In the UK, despite the disclaimer, it was felt that if GMAG advice was not followed and anything untoward occurred, then the courts would probably question the failure to take the advice, especially as the HSE had the power to implement legal action in the first place.

One final limitation of the regulations introduced was that they only referred to the genetic manipulation procedures and not the *use* of the products. Although few regulations in any state covered usage, it was thought by GMAG that the HSW Act would put an obligation on any manufacturer, importer or supplier to ensure safety.[62] To include the use of recombinant DNA molecules in the definition of genetic manipulation could only be done after a further consultation exercise. Nevertheless, GMAG expected that use of recombinant DNA molecules would be notified to them on the basis of it being expedient to use GMAG as a source of advice. GMAG, in keeping with the NIH and recommendations of the European Science Foundation, which had established a Liaison Committee on Recombinant DNA Research, would also ask that anyone transferring products of genetic manipulation abroad, should request assurances that safeguards such as those of the NIH or UK be adopted.

Industrial Confidentiality
The very nature of GMAG's broadly representative membership was to cause a number of difficulties concerning the examination of industrial proposals. Specifically, the problem arose because industry feared that disclosure of information to GMAG, when notifying intent to carry out an experiment, would invalidate any future patents applications. In addition, industry feared the potential loss of industrial secrecy in that six members of GMAG were either management or employees' representatives who might have contacts with rival firms. With increasing interest shown by firms in utilising recombinant DNA techniques, these fears became more urgent. Although GMAG members all had to sign the Official Secrets Act, it was thought that this offered little or no protection in terms of industrial or academic confidentiality as the Act referred to government property. As

a result, the chairman of GMAG established a subcommittee of 14 members to examine the whole question of the confidentiality of proposals.[63] But to complicate matters further, the four trade union representatives, while respecting the need for confidentiality, refused to be bound by any secrecy agreement which could compromise their duty to their members. Nevertheless, pressure was on GMAG to do something in case industries with multinational connections took their research abroad.[64] After consultations with industry, GMAG, also allowing for trade union concerns, was able to introduce a formula for a trial period, initially of six months.

First, any experimenter could ask that certain information be treated as 'sensitive', the chairman responding by deciding whether or not to accept the request and allow the proposal to proceed under the formula. Secondly, all members of GMAG were asked whether any other interests they held would influence their judgement in sensitive cases. Thirdly, those members with conflicting interests would be required to withdraw from discussions on the proposal and would receive no relevant information. Finally, all members who would see the details of the proposed experiment would have signed a confidentiality declaration, to be broken only with the proposer's permission. To supplement this formula, GMAG obtained assurance that disclosure of proposals to GMAG would not in itself constitute prior disclosure within the terms of patent legislation.

Although recognising a diminishing of effectiveness in asking people to withdraw over some proposals, after a year of using the scheme GMAG was able to conclude:

This scheme is not ideal, but it has operated with reasonable goodwill on the part of industry (or others asking for commercial-in-confidence treatment); and GMAG has found it flexible to consider proposals in this way in spite of the diminished number of members present to assess them.[65]

In its investigation of 1978/79, the Select Committee on Science and Technology was to express its unhappiness with the scheme, despite the fact that only a small number of proposals were affected. Taking evidence from members of GMAG (including Sir William Henderson, the second chairman) and Shirley Williams, the Secretary of State for Education and Science, they pressed the point on whether safety assessment would be compromised. Mrs Williams defended the measures by saying: 'I think our compromise is probably the furthest any country has gone in trying to regulate private-sector research in a very advanced field with very great commercial implications.'[66] Acknowledging the arguments it heard in defence of the scheme, the Sub-committee on Genetic Engineering concluded by suggesting it was possibly unsound that such an important advisory group as GMAG should have 'apparently first- and second-class members'.[67]

It must be said that Mrs Williams' comments were more to the point. The UK was very successful relative to other states in providing coverage of industry. Most states operated on a voluntary basis in this respect. Confidentiality had been a recurrent phenomenon in many states, including the US, and whatever the faults of GMAG's compromise when the inspectorate role of the HSE was allowed for, then the UK system was comprehensive. Besides, those likely to be excluded, such as trade union members, were not likely to assess technical details anyhow. Whether they saw the proposal or not they could at least guarantee that it was assessed. By its very nature the problem was difficult and GMAG, itself representing a broad spectrum of interests, had put much effort into obtaining the compromise. In the US, for example, no single body existed with such a range of represent-ation to address such problems. Thus the GMAG approach in this sense had some legitimacy.[68] Not least, GMAG had a good conception of the wider political dimensions to the dis-cussion on confidentiality, and accepted the general under-standing of the time that the UK should not have a regulatory framework which would put workers, including industry, at a competitive disadvantage.

In retrospect, the system appeared to work reasonably smoothly without any real disadvantage. Nevertheless, some industrial spokesmen suggested reform. For example, the Association of the British Pharmaceutical Industry (ABPI), arguing that the UK was unique in its requirements of industry to disclose plans, suggested that this would indeed put British workers at a 'grave disadvantage compared with their colleagues in other countries'. From this the ABPI went on to argue that the technical roles of GMAG and DPAG should be taken over by a new Biohazard Advisory Group, run by the HSE. Policy would be developed by a Biohazard Advisory Committee, analogous to the Advisory Committees on Dangerous Substances and Toxic Substances, already established under the HSC. The chemical giant ICI similarly requested that GMAG be put under a professional group such as the HSE.[69] Criticism of this sort can be seen to reflect two things: first, there was a belief held at the HSE that industry preferred to work with it, partly because of a long association with the Factories Inspectorate; secondly, the ABPI did not allow sufficiently for the whole area of uncertainty and conjecture surrounding recombinant DNA which was quite different from 'dangerous' and 'toxic' substances.[70] It can be argued that there were at that time uncertainties regarding the future use of genetic manipulation techniques in large-scale industrial processes. With the general uncertainty the experience of GMAG might have been invaluable, particularly with its flexibility. As it happened, by the time large-scale industrial involvement became significant, in terms of the imminence of developing products, the overall fears concerning recombinant DNA had declined.[71]

The HSE well understood the techniques of inspection of laboratories and industrial premises, but GMAG, being independent of them, could take a wider view of developments regarding conjectured risks and benefits. These functions were different and usefully separate. GMAG was suited to keeping a more general view of changes in the state of knowledge, and demonstrated this when it subsequently attempted a change in approach to the guidelines. Besides, whatever the complaints of industry, the inconvenience they suffered in complying with GMAG was not great. Indeed, both the ABPI and ICI acknowledged the speed of GMAG's responses to their submission. If the tone of the comments also implied favour for the HSE, it should be pointed out that communications between GMAG and the HSE were very good indeed. HSE representatives attended GMAG meetings and often joint inspections of laboratories were undertaken, notably of

Categories III and IV.[72] Communications were, for example, significantly better than between the RAC and the Advisory Committee to the Director, NIH.

GMAG and the RAC

It was a credit to the success of GMAG that in the US the RAC was to become more similar in its composition. The requisites of legitimacy eventually made their mark and the RAC broadened its membership to include individuals with knowledge of law, public attitudes, public health, occupational health, professional conduct, or related fields, and who comprised at least one-fifth of all the participants.[73]

Public debate was much greater in the US, and this was in part over the question of legislation. As the Federal Interagency Committee discovered, no existing statutes fitted the requirements of comprehensive application of the NIH guidelines to all researchers. In addition, local legislation became an issue. Public debate was facilitated by the relative openness of all the RAC meetings. By contrast, much of the UK machinery operated in secret, although GMAG was a little more open, given that some of the participants reported back to other bodies which they represented.[74] Susan Wright argued that the Official Secrets Act was such that 'not even an illusion of openness characterized the proceedings of the various committees that examined the problem'.[75]

However, the degree of openness was not the only factor affecting the levels of debate. Representation of different values was limited at the national level of decision-making in the US. From the beginning, by contrast, GMAG appeared more legitimate, was innovative and tended to engender more faith. On the contrary, the RAC with its domination by involved scientists appeared more self-serving. Yet most participants in the UK system would probably not have welcomed wider public involvement, as the inclusion of industry in the regulations did indeed make confidentiality a key issue. Trust in the representation, and notable support for GMAG from the trade union movement, negated more widespread concern. Similarly, scientists were not keen on publicity in the UK, perhaps fearing controversy as in the US and, as indicated in the next chapter, France. At two significant conferences, the one in Oxford in July 1975, prior to the Williams Report, and an international one at Wye College, Kent, in 1979, it was planned to exclude the press.[76] Legitimacy in the UK might have been enhanced further if more open discussion had been encouraged. Protecting the public from hazards always appears more credible when it is not just done, but is seen to be done. Despite this, the UK and the US 'packages' of control options became attractive options for other states to model their policies on.

To an extent, a degree of openness did enter the UK methods, when GMAG proposed to revise the UK guidelines. On 9 November 1978, it published in *Nature* a radically different approach to allocating containment, and invited comments, rather like NIH practice.[77] Discussed further in Chapter 6, it was hoped in the new approach to assign where possible theoretically or empirically derived numerical values to the risks involved in each experiment before determining containment. The consultation process was more successful than the earlier exercise involving the HSC document released with the Williams Report. Valuable responses were received and the new system was adopted.

THE SELECT COMMITTEE ON SCIENCE AND TECHNOLOGY

Until late 1978, Parliament's role in the recombinant DNA debate had been limited to some questions tabled on 15 June 1978 by the Labour backbencher, Leo Abse. Eleven questions in all were put to Shirley Williams on safeguard policies regarding genetic manipulation, and he called for 'genuine control regulations'. Abse received a general reassurance that things were under control, but had himself, according to *New Scientist*, initially failed to understand the subtle relationship between GMAG, the HSE and the HSW Act.[78] Abse was not the only one who was unsure of that 'subtle relationship'. It was in December 1978 that the Genetic Engineering Sub-committee of the Select Committee on Science and Technology began taking evidence, much of their investigation, indeed, examining that relationship. The subcommittee had been established in November 1977 but, under the chairmanship of Mr Arthur Palmer, had decided that before commencing their inquiry in full they would hold a seminar on the science and technology involved. Thus on the day Abse tabled his questions, the subcommittee held their seminar at the University of Bristol. Their investigation proper was deferred until after the publication of the HSE notification regulations in August, which in effect meant the next session of Parliament.[79]

The subcommittee declared its interests to be in examining the public policy issues of laboratory and industrial use of genetic manipulation, including the distribution of manipulated organisms, domestically and internationally.[80] During December 1978 and early 1979, the subcommittee undertook an extensive investigation interviewing represent-atives of the key organisations and departments involved, inviting submitted memoranda, and which included five members visiting the US for ten days.[81] The final document was, however, unusual in that the report only covered four pages, but with the minutes and appendices accounting for the remaining 263. The report was an interim measure because of an impending general election in 1979, after which the Conservatives replaced Labour in office, and subsequently disbanded the Select Committee on Science and Technology. The value of the Report itself was therefore somewhat limited. Something of this Conservative attitude had emerged before the general election when the then Shadow Minister for Education and Science, Mark Carlisle, expressed his faith in experts and his hostility towards bodies such as GMAG and DPAG having wider representation, including union and lay representatives.[82] After the election, although no significant changes occurred in the composition of these bodies, or their operation, Mark Carlisle himself delegated the recombinant DNA issue to a junior minister, taking little personal interest, unlike Shirley Williams.

Despite its short length, three conclusions were apparent in the report. First, recognising the difficulties of including industry in the safeguard system, the subcommittee expressed dissatisfaction with the current arrangements. Secondly, they were critical of the DES being the 'lead' department, and would have preferred more involvement of the DHSS, which ran DPAG. Thirdly, they feared UK workers being at an international competitive disadvantage as a result of controls. In a sense, as Roger Lewin observed,[83] the subcommittee can be seen to have modified its concerns. Emphasis altered to a degree from concern over hazards to concern over regulatory hindrance of the work. Perhaps influenced by the CBI, the ABPI, ICI and the AUT, who all wished for administrative changes, the efficiency of the current role of GMAG was questioned. Most witnesses, nevertheless, as late as 1979 did recommend caution and perhaps only a fuller report would have done justice to all the issues. Some of these conclusions have been examined in discussing the UK procedures above.

During the period of the subcommittee investigation, a significant international meeting was held at Wye College, Kent. Organised in part by the Royal Society and a committee of the International Council of Scientific Unions (ICSU), the conference met between 1 and 4 April 1979. Notably, the gathering marked an attempt by many scientists to undo the consequences of their own earlier actions at Asilomar II, with the UK chosen as the venue because it was viewed as 'a small, quiet country where a display of emotion is, to say the least, embarrassing'. These were the words of M.G.P. Stoker in his introduction and welcome to the conference.[84] Criticism at the meeting was most noticeable in coming from Roger Lewin, one of the few journalists eventually to attend, and from Donna Haber of the trade union, ASTMS (see Chapter 5). The meeting itself was rather biased, due to the shared views of the majority of those invited to attend, further reflected in the fear of publicity on the part of the organisers. In so far as the Royal Society was involved in the organisation, it can be said that they tended to see the issues essentially in the light of challenges to the 'freedom of science'. Indeed the President saw one of the two reasons for the questioning of how recombinant DNA work was conducted as 'essentially ideological and [including] quasi-religious objectives'.[85]

The Select Committee hearings and the Wye conference did at least help to focus some media attention on the way by which genetic manipulation was emerging both in the UK and internationally. If a tentative conclusion were to be made, then

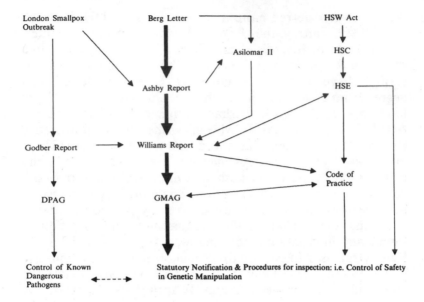

4.1 The development of UK regulatory procedures

the efforts of the parliamentarians were more laudible than the efforts of the scientists at Wye.

THE UK REGULATORY SYSTEM

Figure 4.1 illustrates the way that the main elements of a UK institutional response occurred, based on early expressions of concern. A three-track development was evident based on three initiating factors-the Berg letter, the London smallpox outbreak and the HSW Act. Through a series of government reports and consultation exercises, a system of control developed, covering both known pathogens and conjectured hazards that might emerge from recombinant DNA experiments. Links existed, however, by which both GMAG and DPAG requirements could be applied if, for example, a genetic manipulation experiment involved a known pathogen. From this pattern of development, an operational system emerged involving interactions between a number of bodies,

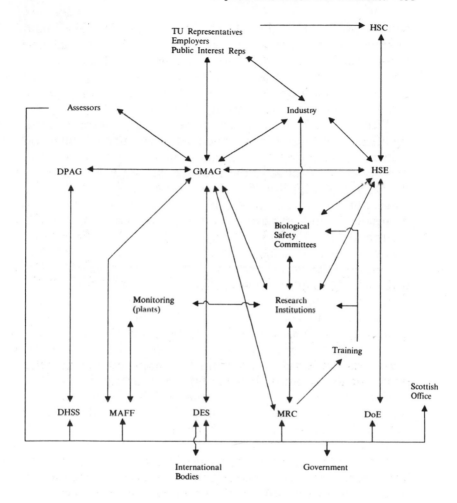

4.2 The UK regulatory framework

Key: DPAG Dangerous Pathogens Advisory Group
      GMAG Genetic Manipulation Advisory Group
      HSE   Health and Safety Executive
      HSC   Health and Safety Commission
      DES   Department of Education and Science
      DoE   Department of Employment
      DHSS Department of Health and Social Security
      MRC   Medical Research Council
      MAFF Ministry of Agriculture, Fisheries and Food

with GMAG and the HSE acting centrally. The main interactions are outlined in Figure 4.2. Both industrial and non-industrial researchers would be monitored in terms of safety by GMAG and the HSE, the latter possessing statutory powers of inspection. At the governmental level, however, the UK situation was not so straightforward.

Within the UK, a total of four departments were involved significantly,[86] a situation which led the Subcommittee on Genetic Engineering to question which should be the 'lead department'. Effectively, the lead has been quickly taken by the DES in its capacity of responsibility for developing science and education. Its leadership covered both the UK domestically and, in terms of official representation, the UK internationally. Criticism of this position of the DES centred on two points. First, the question of conflict of interests arose, in that the DES was charged with promoting research.[87] This undoubtedly had influence in the early unquestioned assumption that the work should continue, the question merely being to determine under what conditions. With the composition of both the Ashby and Williams working parties, it was unlikely to be an assumption challenged by them either. Yet if the view is taken that whatever happened in the UK the work would continue, then the DES can largely defend itself against a bias to promote recombinant DNA work. The only regulation involved was to notify GMAG and the HSE of intended experiments in some detail. From then on both GMAG and the HSE took over, though not forgetting that the DES appointed membership of GMAG. Besides, the HSE was not charged with promoting work. A second criticism, specifically brought up by the subcommittee report, was that perhaps the DHSS would serve better as the lead department given its concern with issues of health and its control over DPAG. A defence of the existing system was in fact made by the Secretary of State for Social Services, who pointed out that the DHSS was more concerned with general health issues rather than research level issues, and that given an outbreak of any epidemic resulting from the work, then it would act. He went on to indicate that there would be difficulties enough in identifying any department that covered all the aspects involved. He thought, therefore, that the distribution of responsibility was about right.[88]

Given that there were diverse departmental interests – the DES with basic research, the DHSS with health, DoE with safety and the Ministry of Agriculture, Fisheries and Food (MAFF) with plant work – then unless a new body was introduced to cover the issue, one factor would be very important – the level and quality of interdepartmental communication. Traditionally, the 'Whitehall system' appointed a lead department in cases where there was not too much interdepartmental rivalry. In the case of genetic manipulation, all the Departments were well briefed about any policy planning and, apart from interdepartmental meetings, acknowledged to have occurred,[89] GMAG itself had assessors from each of the relevant departments in attendance at its meetings. As it happened, the extent of cooperation raised a worrying issue as far as the subcommittee hearings were concerned. A submitted memorandum from the DES referred to what was said in memoranda from other Departments, leading the subcommittee to conclude that a degree of collusion had occurred regarding their evidence. N.T. Hardyman, Under-Secretary at the DES, argued in response that rather than 'collusion' their communication was an attempt to give the subcommittee 'the most helpful and most explicit description we were capable of giving'.[90] The point, therefore, would seem proved, that communication was good between the Departments, at least when it suited them. In so far as the implementation of safeguard measures is concerned, such collusion is desirable when there are a number of departments which must cooperate, whatever the ramifications are regarding the relationship between Parliament and the civil service. The lead Department system again applied at the transnational level with the DES taking the prominent position.

This framework was to remain in place for some years. Time, however, has brought change as perceptions of hazard have modified and as industry and agriculture have become more active in using genetic manipulation within bio-technology. Following the circulation of an HSE consultative document in April 1983 the then favoured government option was implemented, namely the transfer of GMAG to control by the Health and Safety Executive. Part of the logic was that with more departments finding work which they were overseeing could involve genetic manipulation, the issue was less one of simply education and science. GMAG, reborn as the HSE Advisory Committee on Genetic Manipulation (ACGM), would be available to give advice to any requesting department. The move also reflected the growth in experience and stature of the HSE itself in the decade since its inception. Further, although the legal requirement of notification of experiments would continue, these would be dealt with by officials, the committee only examining exceptional proposals. Notification of scale-up proposals would, unfortunately, remain voluntary. By early 1986 the ACGM was working on guidelines for the release of genetically manipulated organisms into the environment. These would operate on a case-by-case basis. Nevertheless, prior to the introduction of the new guidelines, the ACGM approved some environmental release experiments involving 'slightly' engineered versions of viruses used in the biological control of pests. The future can only hold many more proposals for such experiments internationally, leading to new questions of safety. While the US Environmental Protection Agency has moved to extend its legal provisions into this sort of area, it is notable that the UK approach is voluntary as far as environmental release is concerned.[91]

Some other points of interest concerning the UK institutional response will be addressed elsewhere, for example in considering interactions with other states and the European Community. Not least, the House of Lords held an investigation into the impact of a European Commission proposed directive on genetic manipulation.[92]

SUMMARY

Again we can make a limited response to the questions raised in the Introduction, as far as the UK was concerned. In the context of developing control over the safety of genetic manipulation research political constraints were not so evident compared with the biases displayed in the US. The structure of

GMAG was particularly innovative in that it included a wide range of participants representing most of the relevant interest groups. Limitations, in this sense, were more noticeable in the earliest periods of response when the working parties of Ashby and Williams had a narrow science membership. However, the Williams group did consult quite widely. Significantly, the lack of public interest enabled political discourse to arise within the resulting framework rather than outside it, although, as in the US, much of the discussion was about the structure and operation of the adopted framework itself. Further, having a suitable professional inspection service effectively neutralised major criticism about the enforcement of the guidelines, despite disputes from time to time over which department should dominate the overall system of safeguards. Industrial applications of genetic manipulation, in large-scale processes, and environmental release of genetically engineered organisms will require controls in future to be flexible and could lead to further political questions about accountability and acceptable norms.

As with the US, the working parties did not examine a range of control options. They accepted the need for a monitoring service, but advocated a flexible system involving continuous advice being researched and dispensed via a central committee. Thus a typically pragmatic British approach developed, based on the accumulation of case information. In many ways, this outflanks all but one criticism regarding the search for alternative control options. With continuous assessment possible, the strictures and effectiveness of controls could be monitored and, if necessary, changes made. Indeed, GMAG regularly despatched *Notes* to all laboratories and bodies involved, updating procedures and spreading information.[93] The one criticism, however, was that the option of continuing the moratorium was never seriously considered. Nonetheless, it could be said that one real alternative which was investigated was the possibility of applying the procedures for controlling the use of dangerous pathogens to the techniques of genetic manipulation.

The UK response was very much oriented to the HSE and provisions of the HSW Act, but with the parallel use of the new GMAG system. In this respect the UK fared better than the US in that it benefited in having a more suitable existing framework, which itself was relatively new. It was only in 1974, the year of the Berg letter, that the HSW Act was passed, bringing the HSE into existence. Thus, general safety considerations were already centralised, with a national inspection provision. Rather than genetic manipulation engendering a completely novel response, there was a significant element of letting the new HSC and HSE framework show its usefulness in this, a new area, but with some overall pragmatic flexibility retained.

Dangerous pathogens, and consequently the methods of ensuring the safety of laboratory workers and the public, represented a notable area of comparison for genetic manipulation in the UK. This comparison was, however, within the broad field of microbiology, and no systematic effort was made to draw lessons from other technologies which displayed similar risk profiles: low-probability, high-consequence risk. On the inspection side, the HSE did have general expertise regarding techniques of inspection *per se*, and some members of GMAG belonged to organisations with wider interests, such as trade unions or industrial firms. Indeed, some individuals themselves were appointed, as public interest representatives, because of their own wide-ranging knowledge.[94] In sum, despite a lack of systematic comparison, there was at least a useful pool of experience within the implementation framework to call upon.

As I have already suggested, communications patterns are probably best examined at the transnational level and with reference to all relevant groups. Yet with regard to the UK it can be said that given the small 'community' involved, including those who used genetic manipulation, communications were generally good. In international terms, looking ahead, links with Western Europe on the control side were better than with the US, while voluminous press coverage made the US activity familiar to all concerned. On that note, it is appropriate to now turn to the international organisations and the general international impact of the recombinant DNA debate.

5   Towards International Harmony?

NATIONS AND CONTROLS

Biotechnology today is very competitive internationally. This is reflected in the industrial activity of the many firms worldwide which are striving for significant market shares in the exploitation of new chemical factories only visible with the aid of a microscope. Industrial fermenters will provide the habitat for the new biological workforce in miniature to produce a whole new range of products or produce existing products more efficiently. Tentative calculations of the size of markets available for products amenable, in the future, to production processes involving genetic engineering were compiled in 1981. Some examples include an amino acid market of $1703 million, an antibiotics market of $4240 million, an interferon market of $300 million, a pesticides market of $100 million, and steroid and peptide hormone markets totalling $636 million. In 1981 it was thought these items would become producible by genetic methods in 5–10 years.[1] Yet, the competitive aspect of research into the use of genetic manipulation techniques was evident from the time of the Berg letter, when the rush began not only to create safety guidelines but also to embark upon initial research. Scientists around the world were very keen to get in on the act. Publicity about conjectured hazards may have accompanied the Asilomar II conference, but interest in the science was also a consequence as many of the 150 attendees found their appetites whetted. Thus, by the 1980s some 30 states had responded to the calls for safety recommendations[2] while increasingly perceptions of scientists were shifting to fears of unnecessary hold-ups caused by domestic restrictions affecting their ability to compete internationally.

Initially most countries that formally responded to the conjectured risks were simply being cautious. A few countries in addition to the US and the UK embarked upon their own internal discussions about issues surrounding genetic manipulation with a view to developing early policies, while most were prepared to await those outcomes. France, West Germany, Canada and Japan quickly responded, but, like the US and the UK, with an international awareness amongst those individuals and organisations involved. This familiarity was in part fostered at the policy level by the important activities of international organisations. Information, ideas, policy and awareness were not, therefore, constrained by national borders. In the following sections the responses of a selection of countries with leading interests in developing biotechnology will be briefly surveyed, before turning to the international organisations.

France

French reaction, indeed, preceded the Berg letter following a grant request by Philippe Kourilsky, made in June 1974 to the National Centre for Scientific Research (CNRS). Kourilsky had proposed a two-part research programme, the second part involving recombinant DNA techniques. In their judgement, made after the Berg letter, the CNRS approved the grant but with an oral recommendation not to use it for genetic manipulation.[3] In the light of this judgement, Kourilsky contacted the ten or so other scientists in France who were likely to want to undertake such research in the future. Collectively, they wrote to the chief of the *Délégation Générale à la Recherche Scientifique et Technique* (DGRST), the main research body responsible for science coordination between the French research councils, requesting that some sort of control system be set up. Meanwhile, the European Molecular Biology Organisation, discussed below, had become an interested party, and French representatives to that body added to the calls for safety controls. With further statements indicating the potential usefulness of genetic manipulation techniques, the DGRST was quick to respond by asking about 20 individuals to form a committee to continue to address the issues.

Generally, press and public reaction in France was relatively vocal encouraging Kourilsky to provide a lecture on Asilomar and the overall issues at the prestigious Pasteur Institute, which itself became embroiled in internal controversy. A split emerged between the younger and older scientists over the desirability of doing the experiments at all. This led in turn to unofficial votes within the molecular biology department, admittedly in a time of financial tightening, over whether or not to construct a special room for the work. The go-ahead resulted from the department's managing committee, despite 80 per cent of the department, including technicians and junior scientists, being by then against the idea.

Such controversial beginnings in France were followed by a 'convention' being signed between the DGRST and the major research institutions that all genetic manipulation experiments being planned by their staff be submitted to a national committee.[4] By June 1975 the French had a formal two-part national committee. The first part was the Ethical Review Group, charged with investigating the philosophical, legal, moral and ethical issues related to recombinant DNA research, a particularly novel provision compared with other states. Both the US and UK committees were more technical in outlook, although GMAG could, if it wished, raise wider issues over particular experiments proposed. The second part of the French national committee was the Control Commission, which met monthly to review recombinant DNA research proposals and to recommend appropriate safety procedures, similar in function to GMAG. In practice, however, the ethics committee, although novel, was not involved that often. Composed of noted individuals or 'mandarins' as described by John Tooze, himself of considerable importance in certain key international organisations, the committee only became active if the technical committee felt the need to pass a case on.[5] Thus it was the Control Commission which was the main central advisory group in France, a group composed mainly of experts in the field, with four observers representing trade unions and technicians. In addition, local safety committees were to monitor compliance with procedures. In the first two years of the formal system 50 proposals were considered.

Initially the French used the Asilomar II guidelines, followed by those of the US, until they developed their own, drawing on elements of those of the US and the UK. Draft guidelines, with Kourilsky as a co-author, were issued in June 1977 and were finally adopted in December of that year. Because the 'convention' in effect covered industry as well, in order to protect future grants from the DGRST, legislation was felt unnecessary. In more recent years the French government, like many others, has actively promoted the development of biotechnology in the country, by amongst other things encouraging the establishment of centres for co-operation between industry and academia. Such national competition was to cause difficulties in establishing early European Community projects in the new science.

West Germany

Legislation may have been felt unnecessary by the French, but this was not the case with the West Germans. In February 1978, West Germany finally adopted its own guidelines after what has been described as a 'fretful search', which, by March 1977, had produced four drafts.[6] Following Asilomar II, the German Research Association (*Deutsche Forschungsgemein-schaft*) had established a Senate Commission for Safety Questions posed by New Genetic Combinations. It was to advise on the construction of containment laboratories, the possibility of legal actions and international cooperation, as well as drafting guidelines. Initially, the Asilomar guidelines were used, followed by those of the US, with their own guidelines, when produced, again taking account of both the US and UK approaches.

From the start the Senate Commission had non-scientist members, drawing on representatives of industry, trade unions and research-promoting organisations.[7] The Commission was to operate in a similar way to GMAG supervising experiment classification, while research grant applications were accompanied by details of laboratory facilities and training of scientists and Biological Safety Officers. Appropriate as these provisions were, West Germany was to have considerable difficulty in tackling the problem of statutory support for their guidelines, even consulting the UK in their efforts. Despite opposition to legislation from the scientific community and the Central Commission for the Biological Sciences, legislative drafts were produced. It proved, however, to be a complicated task in the context of Federal and state constitutions. By the 1980s the resolution of the problem was still elusive, although at the time of writing West Germany was on the point of introducing laws to cover biotechnology as a whole.[8]

Legislation was one of the difficult tasks internationally, as demonstrated in the controversial responses in numerous countries, including the US which probably had the most international publicity, and would subsequently embroil the European Community. The problems of the legislative approach to safety provisions will be returned to alongside the problems of achieving international harmonisation of policies.

Canada
Canada quickly introduced its own guidelines, which were in turn made known internationally. The Canadian Medical Research Council (CMRC) had taken reponsibility after Asilomar II for the establishment of an *ad hoc* committee to look at the safety issues. A draft report followed, which was widely distributed abroad, before the US guidelines were published. By February 1977, Canadian guidelines were produced which would apply to all research funded by the CMRC and, with their agreement, research funded by the National Research Council and the Council on Research and Health. The guidelines, however, were significantly different from those of the US and UK, in having six levels of physical containment and three biological, and in having wider scope than just genetic manipulation.[9] Responsibility for the implementation of the guidelines was to rest with the individual research institution, and in particular with the principal investigator, while the CMRC was to adopt, again, a GMAG-type role in allocating the categories of containment for individual experiments. In structure the CMRC national committee comprised five laymen, three 'generalists' and four scientists, only one of whom was actually using genetic

manipulation techniques.[10] The Canadian approach, like that of a number of other countries, was to have elements of both the American and British models. But in addition the CMRC took it upon itself to provide for the necessary equipment needed for research institutions to comply with their guidelines. Scientists in many other countries would have welcomed such generosity.

Japan

As of May 1980, Japan had the third highest number of laboratories involved in recombinant DNA research, after the US and UK. It should also be said that as far as the more established field of biotechnology is concerned, Japan undoubtedly leads the world, even if it was relatively slower to take up the application of genetic manipulation in this respect. It is clear, despite this, that Japan has every intention of making up for its initial hesitancy. Japan's dominance of the more traditional aspects of biotechnology has rested upon an historical development of fermentation industries and its ability to utilise 'both engineering know-how in enlarging the scale of efforts and quality control associated with commercialisation'.[11] Nevertheless, Japanese response to the Berg letter was quick and widespread involving many newspapers and journals carrying articles by biologists and critics. In general, one commentator has argued, there was an atmosphere of hostility to genetic manipulation work amongst non-professional readers.[12] Professional scientists responded only a little slower. On 9 September 1974, the Genetics Society of Japan took up the issue at its annual meeting, and prior to Asilomar II it suggested that the Science Council of Japan (SCJ) should study the implications of the research. In January 1975 it was recommended to the President of the SCJ that an *ad hoc* committee be established. The Committee on Plasmid Research resulted.

As in other countries, the feeling amongst the scientists was that the caution called for in the Berg appeal was acceptable, in this case reflected in a poll conducted by the Mitsubishi Life Science Institute.[13] Again, as in other countries, the SCJ Committee on Plasmid Research set up an investigation into the safety procedures and recommended the adoption of elements of both the US and UK policies. They also called upon government to finance safety equipment and restrict funding for recombinant DNA research until a Steering Committee ruled on safety. But the Japanese went further than other countries in widening the scope of investigation to locate the new science in terms of its societal impact. To this end the SCJ established yet another special committee, on Science and Society.

In most states it was the pressure groups that showed the most interest in the consequences for society of the new ability to manipulate the building blocks of life itself. Japan represents an example whereby broad assessments of this kind were sponsored by official agencies. Although in later years formal discussion in the US encompassed more philosophical and ethical considerations, to the extent of a presidential inquiry,[14] it was Japan which adopted this approach as a matter of course quite early on. Principles applied to nuclear energy by the SCJ, namely 'independence', 'democracy' and 'open to public scrutiny' were, it was argued, to be applied to genetic manipulation.[15] In effect this attitude was typical of a Japanese tendency to subject such new developments to inter-disciplinary and interagency investigation. A study group representing eight government ministries was established, with an ultimate objective of producing guidelines, which would in particular emphasise the inclusion of industry. Overall, Japan was most attentive to developments internationally, sending large parties abroad to examine controls in other states, and subsequently entered the ranks of countries with their own guidelines and nationally directed system. Its Steering Group and Advisory Group, combined, included by 1980 a broad range of expertise, both non-scientific as well as scientific.[16]

Some comments on the many national responses to the conjectured hazards of genetic manipulation are now appropriate. Not all countries can be taken individually without adverse repetition and unnecessary detail. For example by 1980, Australia, Belgium, Brazil, Bulgaria, Czechoslovakia, East Germany, Hungary, Ireland, Israel, Italy, Mexico, the Netherlands, New Zealand, Norway, South Africa, Sweden, Switzerland, Taiwan, the USSR and Yugoslavia also embarked upon establishing some form of national committee and applying borrowed or indigenous guidelines. In total some 30 states by this time followed the minimum response of establishing a national advisory committee, with functions ranging through giving technical advice, giving observations on social issues, examining individual experiment protocols, examining legislative possibilities and organising training, to drafting guidelines. However, the countries that have been considered, combined with background reports on the others,[17] lead to certain observations. Not least, it should be noted that the Berg letter and the Asilomar II international meeting had a very great impact worldwide, even in states which at the time had no laboratories actively considering moving towards using recombinant DNA techniques. In part reflecting their functions, the composition of the national committees varied in the range of expertise and interests represented. Thus the following types of response were all evident:

> Adoption of the US package as a whole (local emphasis).
> Adoption of the UK package as a whole (central emphasis).
> Use of the US (detailed) guidelines with a GMAG type committee.
> Use of the US and UK guidelines in combination.
> Development of an indigenous system (perhaps subsequent to one of the other alternatives).

However, over the last decade the international situation was by no means static. Many of the states involved undertook processes of revision for both guidelines and implementation procedures. Not least, the US and UK both revised their guidelines, the RAC was modified by broadening its membership and GMAG was transferred to control by the Health and Safety Executive. Incidentally, the latter move also meant the removal of public interest representatives, to be replaced by more industrially-minded representation.[18] Although some states had guidelines more stringent than others, there was a tendency for them to even out, although invariably with revisions which relaxed the more restrictive guidelines. Yet, in the longer term the issue of safety will have to be considered in an ever-increasing number of countries, stretching across the spectrum of economic development and embracing the field of biotechnology as a whole. As commercial pressures heighten internationally we may need to watch for industry exploiting the differences between national guidelines (where guidelines exist), especially with regard to developing countries that may lack experience in safety policy and its enforcement. International harmonisation of guidelines and the consequences of the growth of international commercial pressures are returned to in the following chapters.

In most cases, enforcement of codes of practice was through central control of funds, which often left industry facing only voluntary control. Legislation to support guidelines, even if only to ensure notification, was rare, although on occasion existing statutes had some utility. However, some Eastern bloc states did claim that they possessed legal sanctions, while many Western states considered the tabling of new legislation. Usually this option was rejected or there was a failure to find agreement between interested parties. The very uncertainty of the risks and the constant revision of perceptions made legislation often seem inflexible.

Overall, information regarding what was happening in other countries was well disseminated between the many national advisory committees, resulting in not dissimilar policy approaches. Awareness of changes in perceptions of hazards, guidelines and the implementation procedures of leading states was commonplace. In particular, the international community followed very closely the early deliberations of US and UK institutions, and the accompanying documentation from the NIH and the UK working parties was well read internationally. Much of the responsibility for international communication of information did, however, lie with the international organisations involved. Some of these organisations took an active role in promoting the exchange of policy thinking and themselves made recommendations, while others fostered and collated supportive technical and scientific information. Not surprisingly, an important view which grew stronger was that there was a need for some form of international harmonisation of approach, a view prominent within the international organisations.

THE INTERNATIONAL ORGANISATIONS

In Chapter 1, much was made of the need to consider the *transnational* features of the biotechnology revolution of the 1970s. The summary of the national responses of states has illustrated the extent of the impact of the calls for caution in using the new techniques, but in order to understand the dynamics of the spread of interest we must consider the role of organisations not confined to the boundaries of individual countries. Science itself is by its nature international and this is reflected in a whole host of specialist international institutions. Many found a role regarding the new biotechnology, especially in compiling and assessing new information. In effect they were critical in linking decision focal points in what was a transnational decision-making system, as far as the development and implementation of guidelines was concerned. The more important of these were, the European Molecular Biology Organisation (EMBO), the European Science Foundation (ESF), the International Council of Scientific Unions (ICSU) and the European Community (EC). There were, however, others such as the World Health Organisation (WHO), the International Association of Microbial Societies (IAMS) and the World Intellectual Property Organisation (WIPO) which became active quite quickly in the area of genetic manipulation.

The European Molecular Biology Organisation

EMBO is an international organisation of scientists from 17 countries, which although non-governmental in structure operates under the auspices and finance of the European Molecular Biology Conference (EMBC), an inter-governmental organisation. It effectively became involved in the recombinant DNA debate when twelve Europeans attending the symposium at Cold Spring Harbor in the US on 7 June 1974 wrote to Sir John Kendrew, the then Secretary-General of both EMBO and the EMBC. They pointed to both the Berg letter of concern and the potential importance of the new techniques. They hoped that EMBO would urgently and carefully consider the problems and in particular provide a special risk laboratory for genetic manipulation.[19] At that time, Kendrew was deeply involved in the establishment of a European research laboratory at Heidelberg, Germany. The proposal was therefore put for the new complex to incorporate a high-containment laboratory for recombinant DNA work. Kendrew took the line that this could only be done if the member governments would provide the necessary finance. The UK, however, opposed the proposal, offering instead the use of the high-containment facilities at Porton Down for European scientists. A counter-argument was noted that a laboratory long-associated with biological warfare would not be popular.[20] With subsequent international developments, pressures for the laboratory grew, leading eventually to a separate building of some 700 square feet at Heidelberg, ironically paid for out of savings on the original estimates for the whole complex. Thus a P4 (US category) laboratory was built for European use.

EMBO, again in response to the Cold Spring Harbor letter, financed five scientists to attend Asilomar II with formal governmental approval at the level of the EMBC. Many other members attended of their own accord. The five were members of an *ad hoc* committee which had been established, by the EMBC in January 1975, to examine the issues. On their return from California the five members presented a report in which three recommendations were made. First, they suggested that the Ashby Report and the Asilomar Conference Statement should serve in Europe as interim guidelines. Secondly, they suggested that EMBO appoint a Standing Advisory Committee on Recombinant DNA Molecules, as part of further elaboration and collaboration in Europe. Thirdly, they added to calls for a European high-containment laboratory.[21]

Established in January 1976, the Standing Advisory Committee, which consisted entirely of scientists interested in the new techniques, met for the first time the following month. Its functions were: to advise, upon request, governments, research councils, national committees, institutes and individual scientists on scientific and technical matters; to explore the possibility of instituting training programmes; and to maintain close liaison with the ESF and other governmental and non-governmental organisations concerned with genetic manipulation. Indeed, over the years this committee was to have considerable influence in Western Europe, directly through providing member scientists with up-to-date information, and indirectly through its advisory capacity in relation to the ESF. EMBO was in part a provider of technical support for a policy function carried out in the ESF (discussed below). A number of training courses were also run.

The provision of up-to-date information operated in two important ways. The first of these followed from the EMBO committee being in a good position to compile information and provide assessment of it. Information was obtained, for example, through members who attended the meetings of other bodies either as observers or participants. An individual of some importance in this respect was John Tooze, the Secretary of EMBO and a member of the Standing Advisory Committee. He also happened to be Secretary to the ICSU, and a member of its genetic engineering committee (COGENE), was a part-time member of the ESF Secretariat and was on the ESF Liaison Committee for Recombinant DNA Research. Tooze also attended the crucial La Jolla meeting of the RAC where the various draft proposals for US guidelines were compared. Indeed, he saw himself as having a key personal role in Europe.[22] In this context he is a fine example of what the Evans model terms as boundary personnel. He was in fact well placed to be of some influence within the transnational community which was addressing recombinant DNA issues, thus linking a number of organisational systems. When contact was less personal, however, the EMBO committee was part of a network of formal and informal communication between individuals and various domestic and international, governmental and non-governmental bodies, a position overall which made its assessments of information of importance.

In its second report, the Standing Advisory Committee produced, for example, a fairly comprehensive analysis of the similarities and differences between the US and UK guidelines. This included assessments of the relative importance in each system of physical and biological containment, and of local and central emphasis.[23] A number of recommendations ensued: that national advisory committees be set up taking into account both the US and UK guidelines; that research protocols be submitted to the national committees, which should in turn specify containment needs; that experiments in the lowest categories allowed to begin immediately; that national advisory groups should have the right to inspect laboratories; that the minimum level of physical containment should be that of the UK, which was higher, rather than the lowest two levels of the US; and that prohibited experiments under the US guidelines should not be undertaken anywhere at that time. Recognising the differences in overall approach in the two sets of guidelines, it was also stressed that combinations of procedures from each should not be used. In addition training was encouraged and, to facilitate international harmonisation, the maintenance of close contact between the national advisory groups was suggested. Analysis of a broader focus was produced by John Tooze, providing summaries of the development of control procedures in Europe for presentation to groups such as the RAC and ICSU. Thus the EMBO committee and Tooze took their roles very seriously both within the context of Europe and worldwide. Compilation, interpretation, analysis and dissemination were all involved.

However, there was a second important means of providing up-to-date information. In line with recommendations made elsewhere, such as in the RAC and the Ashby Report, the EMBO committee argued that 'suitable experiments should be undertaken to assess conjectured risks associated with recombinant DNAs and to pave the way towards eventual adjustment of existing guidelines'.[24] In particular, a risk assessment experiment was proposed by the committee's Swiss chairman, Charles Weissman which would involve co-operation with Ken Murray of the University of Edinburgh, to be performed in the UK and therefore under GMAG oversight. It was intended to use a polyoma virus to which mice were known to be susceptible. The genome of the virus would be incorporated deliberately in a recombinant DNA molecule inserted into *E. coli*, which would then be introduced into the mice. GMAG approved the experiment to be carried out under the maximum containment (Category IV) at Porton Down, Salisbury. As many as 1000 mice were thought necessary for a study lasting some months. Originally proposed in 1977, the experiment was completed in 1980. Risk was also investigated through joint workshops held with the US NIH, one held in 1977 looking at physical containment to reduce risk, while a second in 1978 was convened to assess the risks associated with recombinant DNA experiments involving the genomes of animal, plant and insect viruses. Such risk assessment efforts will be returned to in Chapter 6.

On the whole, communications between EMBO and other organisations were very good, and the advice given was valued by many bodies.

The European Science Foundation
If EMBO was the main European organisation for technical advice, then the European Science Foundation was the main source of policy advice. Membership consists of representatives of 47 academies and research councils across 18 member states. Established in 1974 as a non-governmental organisation the ESF was not, however, devoid of indirect government influence, in as much as governments can influence the policies of their research councils. The organisation grew out of international discussions in the 1960s and 1970s regarding the development of European research. Improved cooperation and coordination on basic science was assumed to be desirable, and the ESF was charged with pursuing these ends. Interestingly, science was seen in its broadest sense and included social science representation.

In April 1975 the ESF established an *ad hoc* Working Group on Genetic Manipulation, composed of 21 biologists, physicists and lawyers. The social, legal and philosophical implications of genetic manipulation were to be examined in the context of giving advice at the European level concerning the responsibilities of scientists and the regulations needed to minimise risk. At a meeting in September 1976, the finalised US guidelines and the Williams Report code of conduct were discussed, leading to a set of recommendations from the committee which were adopted by the ESF assembly the following month.[25] Taking the now familiar view that the US and UK systems should not be intermixed, in part because of the possibility of opting for the lowest common denominators in containment, the ESF recommended the UK approach for

European states. It was argued that the UK code of practice covered all laboratories, and required a slightly higher level of physical containment reflecting greater use of a more tried and trusted method compared with biological containment. Also, more flexibility would be derived from assigning containment to each individual experiment. Perhaps most importantly it was felt that the existing UK legal provisions would have their counterparts in other European states. The ESF further hoped that countries with less experience in genetic manipulation would consult the EMBO Standing Advisory Committee and they specifically requested that the above EMBO comparison between the US and UK guidelines be made. Again, as with all other science organisations, the ESF recommended that the work should continue.

The ESF clearly saw the need for good communications between the various international and national bodies and endorsed the policy of establishing national genetic manipulation committees. For its own part, the ESF established the ESF Liaison Committee, drawing, for its membership, upon the national committees, the EMBO committee, and the research councils of the European countries. As a policy forum the new committee, which met for the first time in March 1977, was to be influential as most national advisory committees were run by their respective research councils. As a further measure of the importance of good communications and the role of the ESF in this respect, meetings of the Liaison Committee were attended by representatives of the US and Canada, individuals who made significant contributions to its business. It was felt that the ESF committee was probably the best forum for first-hand information on all recent developments and decisions.[26] Thus EMBO provided a general focus within Europe for the channelling of technical information, while the ESF committee provided a similar focus for policy information. The EMBO committee in effect provided an important back-up service for that of the ESF. Guideline harmonisation, the relative stringency of guideline options and the need for legislation were all important policy questions to pass through the committee. Of particular importance was the role of the committee in establishing the views of affected states when the

European Commission produced a Draft Directive on recombinant DNA research.

From the point of view of the UK, as an example, the Department of Education and Science (DES) acted as the 'lead' department in international aspects, and therefore kept close contact with both GMAG and MRC involvement in the ESF. Indeed, the DES concurred with the view that the ESF committee represented the most important forum in the development of European responses, and the Liaison Committee itself felt that by 1980 similar safety precautions were in fact 'effective or envisaged in all countries represented'.[27] By the late 1970s the European Commission was forcefully arguing that international harmonisation was essential and that a Directive was required. The ESF was putting the view that this had really been achieved without a Directive. This was, however qualified in that the ESF committee had come to the view that no one set of guidelines or procedures could fit the differing domestic political situations of the member states. Even though the committee was to come to favour revised US guidelines it would, for this reason, not specifically recommend them.

The International Council of Scientific Unions
Despite the overall international awareness and active involvement of the previous organisations they were predominantly European in outlook. This was not the case with the International Council of Scientific Unions (ICSU). Founded in 1919 as the International Research Council, the ICSU took its present name in 1931, and in 1963 adopted new statutes. Its structure includes 18 independent scientific unions with membership from more than 60 national bodies such as research councils or academies. It is thus non-governmental. In general the organisation tries to encourage scientific activity 'for the benefit of mankind' and takes its membership from both the East and the West. It has at times organised major cooperative ventures such as the International Geophysical Year and the International Biological Programme, while on a broader plane it has encouraged international scientific communication, organised international conferences, congresses and symposia, and published journals. Close

cooperation and financial support come from international organisations such as the World Health Organisation and UNESCO.

When scientific activities of a wide-ranging nature arise and the scope of the project is of clear interest to several unions, then the ICSU moves to bring them together to form a Scientific Committee. Many such committees have been formed, one of which was to look at the genetic manipulation issue. The ICSU was in fact quick to act on the developments in recombinant DNA techniques, in part due to the fact that Sir John Kendrew became Secretary General of the ICSU exactly at the time of the Berg letter. Combined with his role in EMBO and the development of the EMBL, he had an interest in getting the ICSU involved. As he put it: 'I really stimulated ICSU to set up a group on a world scale to consider these problems.'[28] Following the recommendation of an earlier *ad hoc* committee, under the chairmanship of W.J. Whelan, a full Scientific Committee was established in October 1976, called the Committee on Genetic Experimentation (COGENE). Seven members came from the unions then concerned, with six others appointed by the ICSU executive. Whelan became chairman with John Tooze as secretary. During the next few years COGENE was to increase in its international influence.

COGENE was designed to serve as a source of advice for governments, government agencies, scientific groups and individuals. Safeguards, containment facilities, training and scientific exchange were all to be monitored and encouraged. Its particular advantage over other organisations fulfilling not dissimilar roles was its global scope, making it in turn an attractive source of advice for other global institutions such as the WHO. COGENE established a number of working parties charged respectively with analysing existing guidelines, sponsoring and gathering information on risk assessment, and studying the requirements of training in safely using genetic manipulation. It was, however, relatively late in the debate when the investigations of the working parties made their impact.[29] Nevertheless, there appeared to be a broad philosophy underlying COGENE's approach to the issue, centred on conserving the interests of scientists. This was especially the case when the criticisms of genetic manipulation became more vocal and calls for legislation ensued. Judgement on this is necessarily subjective but supported by a number of indicators, partly relating to the committee's relationship with the press and its wish to avoid publicity. Yoxen was to describe it thus: 'COGENE became, in effect, a pressure group for minimal regulation of this research and its members exploited all their connections in governments around the world to get the message across.'[30] Although the reports produced by the working groups are impressive in terms of the scope of the international assessment involved, it should be said that COGENE tended to keep its substantive inquiry to relatively technical areas. The group examining guidelines had embarked on a questionnaire survey of all the ICSU member states producing a report in 1980, while the risk assessment group, using information from questionnaires and analyses of workshops held by other organisations, came to this conclusion in 1978: 'no risk unique to recombinant DNA research has been identified.'[31] In addition, over the years, much training in the use of genetic manipulation has been organised for countries where such training was not available (Brazil, India, Yugoslavia, South Africa, Hong Kong and Costa Rica).[32] Broader issues, however, were downplayed as the new science was encouraged to flourish. This aspect of the approach taken was most noticeable when COGENE decided to hold an international meeting addressed to genetic manipulation.

By April 1978 a proposal had been put forward to bring together, in an international gathering, scientists, research directors, legislators, lawyers and public health experts. It was to be a multipurpose gathering to assess the contemporary status of the new science. This meeting, held in April 1979 at Wye College, Kent, was both important and controversial. Ironically the criticism grew out of an attempt to mute adverse press coverage which backfired. The organisers had hoped that the meeting could take place in 'a cool, uncharged atmosphere, at least until the conference had assessed the situation and come to some conclusions'.[33] The UK was selected on the basis of its quietness, and London was excluded, it was said, in favour of a venue which would provide more opportunity for 'informal' contact.[34] A letter from the organising committee, dated 12 March 1978, to all participants had given the impression that no members of the press would be present and it requested attendees not to contribute to reviews for publication other than the official proceedings.[35] Belated changes were to be forthcoming as fears arose about the possible misinterpretation of these actions. In response to a request from a writer for *New Scientist*, the organising committee had agreed to invite three members of the Association of British Science Writers, provided that they agreed not to record or despatch reports directly from the meeting, in a fashion similar to the Asilomar II requirements. The remaining ban was only lifted the day before the meeting commenced – a Sunday.

Strong criticism ensued from the press. A *Nature* editorial prior to the meeting suggested:

the committee has already set the tone – the discussions are too sensitive to be widely disseminated except in an official version. Others may conclude – perhaps quite wrongly – that there are other things to hide.[36]

At the meeting itself Roger Lewin of *New Scientist*, in a presentation to the gathered assembly, lambasted the attempt at an off-the-record meeting and the approach to press coverage. He also suggested that the choice of venue was too remote, adding cause for suspicion, especially given the aims of the conference, which he argued included the presentation of views and evidence to show risks as much less than orginally postulated. Regarding the question of risk, *Nature* was to observe that change of heart as follows: 'To the outside world the views that risks are negligible went through on the nod.'[37]

To be fair to COGENE they always intended to publish the proceedings of the meeting as quickly as possible, and to hold a press conference after the event. Yet the fact remains that of the 143 participants attending a very important international conference, only three were from the press, and only the UK press at that. It is generally acknowledged that the approach to the press was bungled. In hindsight, Whelan, himself, argued that the correct approach was taken when the organisers realised that their call for an informal meeting was being mis-understood.[38]

Perhaps a more subtle weakness in the conference was the absence of noted scientists who had been critical of the claims regarding the safety of the work. Criticism was left to a few individuals such as Donna Haber of the UK trade union ASTMS. She feared the hasty dismantling of the guidelines, and accused scientists of 'overreacting' much as they had accused the public in its response to their early expression of concern. She went on to point out that: 'the work really hasn't been stopped and it seems to me that it's gone on and gone on very well. I think that a meeting like this is not a way to reassure the public.'[39]

In summary, COGENE became an important actor regarding the issues surrounding genetic manipulation. Over time it was, however, to change its emphasis away from guidelines and risk assessment towards stressing the benefits of the science and promoting training. Biases there may have been in its overall attitude but not, perhaps, too surprising within the framework of an organisation primarily active in the promotion of science *per se*.

THE EUROPEAN COMMUNITY

Of all the issues to be addressed by the various international organisations, it was the proposed action of an international governmental organisation which was to be considered by them as most controversial. The organisation concerned was the European Community and the proposal was for a Directive on genetic manipulation which would have compulsory legislative effects in the member states. Directives have the status of being binding on the members, but leave them to choose the means of execution.[40]

Moves first began in this direction in January 1977, two and a half years after the Berg letter and six months after the US guidelines were issued. Part of the Commission's role is to put forward proposals for Community legislation, and the idea for this Directive emanated from the Research, Science and Education branch of the Commission (Directorate General XII). The objective was to harmonise the precautions taken in the member states regarding recombinant DNA research. In early 1976 there was not too much opposition to a Directive, provided it was not too specific or detailed. This was to change, however, once the draft was produced. While it was evident that most people involved saw that the Commission possessed the authority both to hasten the harmonisation that was felt to be needed, and to incorporate the industrial private sector within a common framework for all recombinant DNA research, the timing worried many. The Commission was entering the field late in the debate, and this could all too easily create resentment amongst the researchers, now that perceptions of the risks appeared to be ameliorating.[41]

In late February 1977, the issue gathered momentum as it was discussed by the Medical Research Committee, a subcommittee of the Commission's Scientific and Technical Research Committee (CREST). Yet no firm conclusions were reached at that time, and it was not until early 1979 that the published Draft Directive was sent formally to member states for their consideration. Indeed, this time-lag represented one of the major criticisms. Many groups argued that the European Community operated too slowly for the sort of exercise it was contemplating in this instance. Nevertheless, a number of drafts were produced, drawing on the advice and experience of groups such as EMBO, the ESF and, within the Commission, the Medical Research Committee and various *ad hoc* groups of experts.[42]

The final published draft outlined its underlying motivations. Predominant was the wish to avoid discrepancies in levels of safety arising out of some states possessing suitable statutory provisions, while others did not. It was also hoped that discrepancies could be avoided in the controls imposed on publicly funded laboratories and industrial laboratories. The draft accepted that the existence of expensive protection devices at the physical and biological levels was evidence of the seriousness of the conjectured hazard, and that consequently the Commission wished to see these measures as effective. The question of risk was taken a little further in that it was argued that as more and more institutions were using the techniques, then the total level of risk was increasing commensurate with this. Further, risk was also seen to be 'transnational', because viruses and bacteria recognise no national frontiers. From this latter observation it was suggested that there would be a reduction to a certain extent of the liberty of states to define and follow independent policies. Supporting this they argued: 'agreements and ... guarantees can best be generated through legal dispositions, taken in each country, which are based upon a core of principles adopted in common.'[43]

Legislation was really the nub of the controversy here. The draft acknowledged that the case of recombinant DNA provided an opportunity to test the possibilities of compatibility between legislation and the development of modern technologies. This was seen as part of the wider objective of protecting man against his own achievements, and particularly the long-term influences on society and the environment of the applications of modern biology. Of course, the counter-argument spotlighted the rapid pace of the progress in the new technology, the consequent increase in fundamental knowledge, and the inability of the legislative process to adapt expeditiously to such changes. Critics held to this despite the Draft Directive's inclusion of a provision for review of the Directive and revisions if necessary at least every two years. This could imply effectively a two-year time-lag.

The scope of the proposed legislation was such that even the UK, upon whose model much of the Draft Directive appeared to be based, would require legislative changes. Central notification of all experiments involving recombinant DNA molecules was called for through the despatch of a detailed experiment protocol. All but the least hazardous would require prior approval from the 'national authority', decisions upon which would have to be made within ninety days. The changes the UK would need rested on the provisions that prior approval was required (rather than simply notification), both the manufacturing of recombinant DNA molecules and the *use* of such molecules were covered, and the Draft Directive gave the national authority powers of revocation of previous authorisation. Indeed, within the UK, the House of Lords Select Committee on the European Communities held hearings on the Commission's proposals, concluding them to be too restrictive and ill-suited to the rapidly developing techniques. The Select Committee suggested that a Recommendation from the Council would be preferable to a Directive.[44]

Given the extent of influence of the UK system on the Draft Directive it is worth looking more at the response to the proposals within the UK as a measure of the controversy. Not least at this time GMAG was on the point of changing to a quite different mode of risk assessment. While DG XII was examining the issues, the DES was the lead Department for the UK and it argued for a Community Recommendation. UK industry was even more critical, as represented by the Confederation of British Industry (CBI). They forcefully opposed the whole idea of a Directive, although recognising the need for international harmonisation. Industry saw the Directive as out of date and argued that it only paid lip-service to the needs of flexible revision. In particular, they cited their relationship with Commission officials leading up to the draft as evidence of the latter's unwillingness to take account of scientific progress.[45] Further, the CBI criticised the limits of having harmonisation across the membership of the European Community, if non-members were not brought into the agreement. The CBI felt that this could hinder European states, should inflexibility be evident given the likely decline in perceptions of risk. There is, however, another side to the industrial application of genetic manipulation, the views of industrial workers as put forward by their unions.

By contrast to the CBI, the UK Trade Unions Congress welcomed the Draft Directive, but with a wish to extend harmonisation beyond the Community. They even suggested a further proposal that all workers involved carried a card stating that they worked in a genetic manipulation laboratory, in the event of illness arising from a laboratory accident. GMAG, as far as the TUC were concerned, was a most suitable model for the content of the Draft Directive, especially with its trade union representation.[46]

In the event, the final result of two years of discussion on the document was that it was withdrawn in favour of a Recommendation. It should be said that the UK was not the only critic. France and Denmark were noticeable in their opposition. Nevertheless, even before a Recommendation resulted, there were further meetings including a public workshop, arranged by the Economic and Social Committee. By 1981, the main provision of the Recommendation was simply the registration of recombinant DNA work, but with prudency reflected in a proposal for at least an annual review of the need for harmonisation.[47] Thus, like most legislative attempts within states, the European Community's effort was unsuccessful. But the Community's interest did not end here. In more recent years there have been efforts to draw up a set of regulations to apply to the biotechnology industry as a whole with provisions for issues such as the release of genetically engineered organisms into the environment. Eighteen months had been spent in negotiations between European civil servants and the biotechnology industry in an attempt to set a European standard before individual countries implemented their own policies. A solution has not been forthcoming. Since December 1984 a committee of 15 industrialists have been examining the question of European biotechnology and have concluded that no new laws or regulations are necessary for most activities except the environmental release of genetically engineered organisms. The Commission has argued that wider provisions may also be needed but has foundered in disagreements on positive proposals between the many Directorates now involved in some aspect of biotechnology. In the meantime, individual member states have begun to follow their own routes. West Germany and Denmark have drawn up laws while other countries are reported as also having started the process.[48]

It would, however be misleading only to examine the European Community in terms of its protracted efforts to regulate genetic manipulation and biotechnology. It also became active in promotion of the research.

The European Community first considered a programme of research and development which included genetic manipulation after a symposium in 1976, given further attention by the Commission in July 1977, in a document submitted to CREST.[49] From this, moves began towards possible Community action, against a background of generally attempting to foster something of a common policy in the field of science and technology. The Commission asked that two studies be made, one looking at genetic manipulation while the other addressed the more established area of enzyme technology.[50] On the basis of these studies, emphasis was put on the importance of developments in biomolecular engineering for agriculture and industry, and consequently the need to promote research. In particular, it had been shown that Europe was not as a whole cooperating to match the potential of Japan (notably in enzyme technology) and the US. One criticism made was that scientists in Europe had more contact with scientists in the US than in other European countries! In addition a statistic of note was that Japan then had 4000 PhD biotechnologists in comparison with about 200 in France.[51]

Despite the powerful logic of a need to coordinate research efforts to compete globally in the rapidly growing and diversifying biotechnologies, the proposed European Community projects, in the form of a five-year plan, met with problems of competition at a different level. National research imperatives, of some of the member states, themselves came to be seen as in competition with those of other member states. Initially it was suggested that 26 million European Units of Account (or approximately £16 million) be divided between nine states over five years to finance six projects. Three of these were of direct relevance to genetic manipulation. Although broadly backed by industry, objections were raised after the proposals were subjected to national consultation.

In particular the French and Germans raised objections, the former wanting more to be allocated for education (50 per cent) while the latter wanted the projects to be reduced in number to two.[52] Part of the problem was a reluctance to duplicate potential national research. Eventually, a compromise was worked out by CREST whereby the number of projects was reduced to four, eliminating areas with likely immediate medical or industrial application, and the budget was trimmed to 15 million EUA. Twenty per cent of this would go to education in response to the French. However, even this was not acceptable at the Council research committee level, nor was a further suggestion of allocating only 11.8 million EUA. Both the French and Germans were insistent in their demands. Britain would accept the CREST compromise, and all other member states would accept the original proposal. It seems that at that time a key problem was that states with strong domestic research programmes and industrial interest were reluctant to share commercially-sensitive information. By 1983 only 8 million ECUs (European Currency Units – formerly EUAs: less than £5 million) had been committed to the further weakened biomolecular engineering programme, although with further expenditure to come.[53]

Time has eventually brought progress in European cooperation in biotechnology, as efforts for a further four-year plan have been devised. Policy has really moved more towards the promotion of an underlying research infrastructure upon which European biotechnology can thrive. In late 1984 the European Council of Ministers approved a plan, given governmental go-ahead the following March, for 55 million ECUs (*c.* £35 million) to be put into fostering transnational biotechnology research and training over the period 1985–89. Support was to be given to a broader range of biotechnological areas than in the first biomolecular engineering plan, although still including genetic manipulation. The aim of the Commission had become the creation of a series of interlinking biotechnology research centres across Europe, on the basis of awarding research contracts and encouraging staff exchanges. Special priorities were given to the application of information technologies in biotechnology, through the use of databanks and advanced computer software, and the involvement of industry which could also compete for grants. It is of some interest here that studies on new ways of assessing risk were also to be included, especially regarding the more recent issue of releasing genetically engineered organisms into the environment.[54]

The European Community has therefore been of some significance in its input into the transnational and international discussion of genetic manipulation. Notably, the issues surrounding the Draft Directive were symptomatic of efforts to establish a role for legislative provisions in dealing with the question of potential risk. Close communication was evident as the debate unfolded between the EC and other bodies, both domestic and international, even if there were noticeable differences of opinion involved.

SUMMARY

Taken together, the range of responses of individual countries and international organisations, illustrate the international impact of the recombinant DNA debate. Selectivity has, however, been necessary in the choice of countries and organisations outlined in this chapter. Other countries, in addition to those above, have had their share of debate, for instance the Netherlands and Australia, while more and more organisations have become involved. Although, some of the Eastern bloc countries provide the only examples of legally enforceable guidelines and sanctions they were, perhaps not too surprisingly, able to avoid vocal controversy. The World Health Organisation has brought the issues surrounding genetic manipulation into the wider discussion of dangerous pathogens from all sources, taking advice at times from the likes of COGENE. The United Nations Industrial Development Organisation has attempted to extend bio-technology expertise into the developing world, while the OECD has fostered awareness of the promise of new biotechnology developments for future modernisation.[55]

Given the concentration on the early decision-making over the coming to terms with conjectured hazards in genetic manipulation, the organisations outlined in this chapter were the most important. As far as information exchange and coordination were concerned, the international organisations were more important than most individual states, and as within many domestic systems, there was something of a division of labour between them. Some organisations specialised in the coordination and dissemination of technical information, such as EMBO and the ICSU, while others focused more upon the coordination or making of policy, such as the ESF, the European Community, and to an extent the WHO. These divisions must be seen as a little arbitrary as some of the technical organisations also displayed policy biases. It could be said that globally a three-way relationship was apparent between the US, the UK, and more collectively continental Western Europe. Many states awaited the outcomes of deliberations and policy choice in the UK and the US, by which time harmonisation imperatives were stronger.

European organisations were conscious of this and were quite influential in the responses of European states. In general, communications between the three points of this 'triangle' were very good.

Yet, it should be emphasised that the predominant international characteristic was that the decision-making occurred within a transnational framework of common knowledge and communication. Scientists and decision-makers, who of course were often the same people playing different roles, were well aware of the international nature of the safety issues. Indeed, the debates about the need for controls, the need for legislation, and the levels of risk involved (not excluding more philosophical and ethical issues) were all enacted with this international awareness. In the following chapters, which investigate the way the substantive issues were tackled in this framework, the transnational nature of the politics involved will be apparent. It will also be apparent that cracks in the trend towards harmony in guidelines would be revealed, as imperatives for national success in exploiting the new technology became more manifest.

6 The Transnational Politics of Risk-Benefit Assessment

How the message of inheritance is passed from one generation to the next was discovered in 1953, and it is the adventure story of the twentieth century. I suppose the moment of drama is the autumn of 1951, when a young man in his twenties, James Watson, arrives in Cambridge and teams up with a man of thirty-five, Francis Crick, to decipher the structure of deoxyribonucleic acid, DNA for short.[1]

This is how Bronowski, in his companion book to his monumental television series, *The Ascent of Man*, described a key leap forward in human endeavour to understand the world in which we live. The discovery of techniques for man to *manipulate* DNA in the laboratory, for his own purposes, must surely represent a further significant leap. In a wider political and social context, however, there was more than a 'moment of drama' to their revelation and subsequent development. James Watson was again to play a central role and in 1979 was to suggest: 'our national leaders should announce that they will help push DNA research as fast as our national and corporate treasuries can permit.'[2] There is no doubt that the new techniques caused tremendous excitement amongst scientists in related fields, particularly once the 'moratorium' was lifted and the work could begin in earnest. However, what also became clear was that despite the earlier expressions of concern, many scientists feared unnecessary hold-ups in the subsequent exploitation of genetic manipulation. It became apparent that groups of scientists went on the defensive against what they perceived to be increasing bureaucratic involvement. An international phenomenon, it was most acute when legislation loomed.

With the strength of feeling amongst scientists, who wanted to embark upon using genetic manipulation, and the corresponding conjectures about potential hazards, certain issues came to the fore. The first of these revolved around the degree of risk that might exist for any type of experiment. From the Berg letter itself, there was a call for the assessment of risks. As the genetic manipulation debate grew in prominence, the international efforts focused upon risk assessment also increased. Not least, the results of risk assessment were used by some groups, wherever possible, to neutralise calls for legislation. With time the risk assessment exercises became more complex, involving the interpretation of new data and reconjectures of hazard drawn from greater conceptual understanding of microbiological processes. These efforts were not without controversy in themselves, given the problems of applying 'rational' analytical techniques, but additionally some exercises raised serious questions regarding the presentation of the results to the world at large.

Arguably risk assessment is too narrowly addressed in many technological activities, a situation true also of genetic manipulation and other recent biotechnology developments. Risk assessment on its own invokes no balancing between the potentially undesirable possibilities of using a technology and the potential rewards. In its strict sense, risk assessment simply attempts to define the probability of the undesirable consequence occurring, given certain levels of use of the technology and safety provisions. Although the broader phenomenon of 'risk-benefit assessment' may not have been given much attention by formal decision-makers in positions of responsibility, this does not mean that it did not occur. In one sense, the idea of 'social accountability' when examining any technological activity implies that users of that technology should face the scrutiny of their larger community. If the technology in question brings possible hazards, then its use will only acquire social respectability if the risks are deemed acceptable *given the likely benefits*. This statement, of course, does not preclude a view that no benefit is worth the particular risk associated with it. The point is that much of the politics that surrounds debates on the impact of technology on society reflects differing perceptions of how such equations should be balanced. Outside the tight frameworks of cost-benefit economics, rationality does not provide easy solutions to these problems. In this sense the genetic manipulation debate involved a broad, politicised process of risk-benefit assessment where legitimised balances between safety measures and utilisation of the new knowledge were searched for by the different participants. Linked to the idea of the assessment of risks in relation to benefits, in this sense, is the invocation of legislated restrictions. The problem in the case of genetic manipulation was that the great uncertainties regarding what risks were involved led to further uncertainty as to whether or not legislation was appropriate. Again, different groups held different views.

All of these issues, risk assessment, social accountability, legislation and the ensuring of guideline compliance, along with others, can be considered by reference to transnational relationships. They are also, of course, related to each other.

RISKS AND BENEFITS: PROCESSES OF ASSESSMENT

Risk Assessment

Formal risk assessments were difficult to put into practice in the case of genetic manipulation. The real problem all along was that the risks postulated were drawn from conjectures about how inserted foreign DNA might behave when placed in certain host–vector systems and what the consequences might be if such modified organisms escaped confinement. It was primarily fear of the *unknown* which predicted the commendable caution of the international science community, rather than collected empirical evidence. But the conjectured risks were nevertheless drawn from deductions, understandable assumptions, and reasoned hypotheses. Further, there was evidence to show that tumours could, at least in primates, be caused by viral DNA, the fear being that isolated random segments of DNA could give rise to novel pathogens of a similar nature. As more became known about genetic processes, through *using* genetic manipulation, risks were conceptually reassessed. But the changing perceptions were not without prejudices.

Sketchy as the early conjectured hazards were, they still represented a particular type of perception of risk. The worst conjectured hazards would have been reasonably described as disastrous had they occurred. A cancer epidemic carried forth by a genetically engineered variety of *E. coli*, or *E. coli*-producing proteins otherwise alien and harmful to humans, were two feared disasters. Some perceived hazards could arise from experimenters genetically manipulating substances already known, or thought to be harmful, such as SV40 (as with the proposed Berg experiment which started the debate) or something like botulinus. Alternatively, the 'shotgun' experiment might give rise to the expression of otherwise dormant segments of DNA, including sections of carcinogenic viruses possibly carried in human chromosomes. Novel toxins or the introduction of certain hormones into people via some host–vector system might have further unknown results. Novel proteins being introduced into experimenters' bodies might trigger autoimmune reactions from their body's defences, if the antibodies responding to that protein were also to attack a similar human protein. Bacteria readily destroyed by antibiotics might acquire new resistances leading to problems of their control. Lastly, and more widely still, animal and plant populations might also find themselves at risk from other experiments.[3]

These perceived *low-probability, high-consequence* risks would be difficult to evaluate. Yet a central concern was the development of safeguards applicable to different types of experiment in relation to some estimate of risk. The estimates of risk were, on the whole, based on the differences between prokaryotes and eukaryotes and the view that risks diminished in relation to the evolutionary distance of the source DNA from man. However, across the international range of guideline approaches, different degrees of emphasis were put on the risks associated with particular types of experiments in relation to the current wisdom. The point is not to delve into the details of such differences of conceptualisation and assessment, but to indicate the consequences of, on the one hand, the perceived need to make some sort of assessment of risk, and, on the other, the requirement that uncertain procedures of assessment and categorisation should be seen as legitimate.

The process of risk assessment took place in many forums. In some ways even the writing of the Singer–Söll and Berg letters were part of that process. Formal assessment exercises followed suit, including various domestic and international investigations, and it was not surprising to find appeals for 'common sense' and rationality becoming synonymous. After all, scientists are thoroughly familiar with 'rational science' where chains of logic are deduced from experimental hypotheses and tested against empirical evidence. They wished to move towards this familiar framework and away from the dependence simply on conjectural insights. What is more, in retrospect some scientists, previously involved in the calls for caution, argued that there had been, in fact, little or no evidence to support that initial concern. The real dilemma, however, was that there was insufficient evidence at that time for the task of formal and rational risk assessment, where rationality involves comprehensive knowledge of alternatives and probabilities of outcomes upon which to base choice. The simpler option had been to base levels of risk on the proximity of the donor DNA to man.

A reciprocal to the argument that more evidence should have preceded the calls for caution is the argument that cautionary safeguards should not be relaxed on the basis of insufficient evidence or evidence given too much weighting. For some years it was the case that the risk assessment process was itself largely conjectural, but increasingly supported by new theories and models of genetic functions. Empirical evidence was never in abundance. It is worth, however, considering the way that risk assessment developed and its international impact as, not least, both the RAC in the United States and GMAG in the United Kingdom were active in transnational processes of risk assessment. The charter of the RAC and the terms of reference of GMAG both explicitly required these bodies to assess risks or hazards in relation to safety precautions. For example, between revisions of the US guidelines, the RAC sponsored risk assessment experiments and investigated potential host–vector systems, while GMAG eventually adopted a new categorisation system more closely related to the technical risk assessment procedures used in areas such as the nuclear and chemical industries.[4]

In 1982, Sheldon Krimsky (both a non-scientist representative on the RAC and a lobbyist within the Coalition for Responsible Genetic Research) produced what is undoubtedly a detailed and informed analysis of the varied approaches to risk assessment adopted at different institutional levels within the United States.[5] His analysis is instructive in terms of the US policy process involved and the use of new information on the part of scientists. In particular, he examined the impact of a workshop held to investigate the potential of converting *E. coli* K–12 into a pathogenic organism.[6] It was held in Falmouth, Massachusetts in June 1977, and was influential in leading to a reduction in strength of the US guidelines. However, some controversy was involved. The meeting had been organised under the auspices of the US National Institute of Allergy and Infectious Disease (NIAID) and the National Institutes of Health, with the intention of including scientists from wider fields than those who were using recombinant DNA techniques, and who would be experts qualified to discuss infectious diseases.[7] Two early reports of this workshop both emphasised its unanimous conclusion that it was 'virtually impossible to convert *E. coli* K–12 into a pathogen of epidemic consequence by insertion of random bits of eukaryotic DNA'.[8] A report by a COGENE representative, A.M. Skalka, and a letter by S.L. Gorbach, the workshop chairman, to the Director, NIH received early and widespread attention and provided support for those wishing to relax precautions. Krimsky in retrospect has raised many questions of the accuracy of Gorbach's letter which had given a strong impression that *E. coli* could not be a vehicle for hazard. By breaking the arguments down, Krimsky addressed the logic of the statements made by Gorbach and indeed the accuracy of his summary.[9] Some points are of note: the technical interpretations of results were questioned to some degree; the argument which was produced by Gorbach related to pathogenicity comparable with *existing E. coli* pathogens, and therefore precluded *new E. coli* pathogens; it was left unresolved whether *E. coli* K–12 carrying genetic implants could pass its genetic information to natural indigenous organisms within the human gut, should it successfully escape confinement. The point is that if such technically-based questions of definition, scope of inquiry and extensions of investigation were shielded behind a veil of strong assertions of safety, then risk assessment of this nature has questionable legitimacy. Krimsky has shown that the logic explicit or implicit in the assumptions and conclusions was weak. But the whole endeavour was presented as authoritative and the 'results' were rapidly and internationally disseminated in the context of Gorbach's letter and Skalka's report for COGENE.

Following Falmouth, an important international meeting was held between 26–28 January 1978 at Ascot, England. Sponsored by the NIH and the European Molecular Biology Organisation, it was reported in the US *Federal Register*.[10] Scientists from 27 countries attended as relevant experts rather than as representatives of governments or policy making groups. Expertise included was very broad, embracing specialists from a whole host of areas, including virology, infectious disease, public health, and animal and plant diseases. Genetic manipulators might inadvertently create new diseases, but these were the people who understood known diseases and their causes. Only five of the participants were in fact actively engaged in recombinant DNA work. The results of this international investigation were, like the Falmouth workshop, influential, not least in the relaxation of US guidelines in 1978. In particular, it was observed from a study of the mechanisms necessary to transfer viral DNA (inserted into *E. coli*) through bacterial replication, that such a transfer would be unlikely. Even if, in a worst-case example, the viral DNA could become established in natural 'wild-type' *E. coli*, and was scattered in abundance within many people or animals, it was considered impossible that any virus particles would be created by the bacterium as it lacked the appropriate enzyme systems found in vertebrates. From this limited ability merely to gain access to host cells, but not to have the DNA replicated in *E. coli*, the report concluded that the containment should be no more than that required for the same virus involved in non-recombinant DNA work.

The European Science Foundation endorsed the recommendations of the Ascot workshop, emphasising the

Report's reference to the need to ensure proper training of personnel as a safety feature, and COGENE considered the workshop a successful risk assessment exercise.[11] Never criticised like the Falmouth exercises, the NIH/EMBO joint effort was laudible. However, it should be said that it was some five years after the first expressions of concern. Both workshops were to have influence on guidelines revisions internationally. Following Falmouth, the revision of US guidelines was to take a year, but amid considerable debate reflecting widely differing viewpoints between practitioners and other interest groups. Much of this was to occur against impending US legislation and reports of guideline violations, discussed below. In the near future a conceptually different scheme for guidelines would be developed in the UK.

In the US, much of the proposed revision centred on the greater faith in the restricted ability of *E. coli* K-12 to become pathogenic after DNA insertions. Detailed proposals originating within the RAC were passed to the Director, NIH and published for public comment in the *Federal Register*,[12] and were followed by a public meeting of the Director's Advisory Committee, convened on 15 December 1977. Interested groups made statements and all the correspondence received since the publication of the proposals was available at the meeting. Ironically, however, the proposed revisions were not to be introduced to the planned timescale. Seven months after the meeting, a further set of proposals was published, this time including a chart comparing the original guidelines, the 1977 proposals and the latest considerations. Despite many criticisms of the first set of proposals and the procedures for making revisions, the 1978 proposals were even more lax, by now being subsequent to the Ascot workshop.[13] A further impetus to downward revision was an explicit comparison with guidelines from other states, noting that some experiments banned in the US were allowed abroad, or at least were subject to lesser controls. No reference was made in the *Federal Register*, however, to requirements in other states which were *more* stringent than those of the US. Many countries were well behind both the US and the UK in establishing initial guidelines, never mind revisions.

After a one-day hearing on 15 September 1978, the revised

guidelines were published on 29 December to come into force on 2 January 1979. Krimsky, however, has taken his analysis of the presentation of empirical work further, in exploring the build-up to still more revisions in 1980, in particular emphasising the role of Wallace Rowe, a member of the RAC. Rowe had consistently argued for extensive relaxation of the guidelines and in 1979 led calls for the exemption of most *E. coli* K-12 experiments from any national guidelines at all.[14] His proposals, which were to be supported by many scientists, including Paul Berg and Stanley Cohen, were seriously considered by an RAC working party, but were opposed by other respected scientists. Rowe had marshalled as much 'evidence', including the results of his own work, as he could find that work with *E. coli* K-12 was safe. An interesting misuse of some evidence, however, was noted by Roy Curtiss. Curtiss observed that much of the evidence gathered related to biologically enfeebled laboratory strains of *E. coli* K-12 (an EK2 system) and not to the wild-type strain of *E. coli* K-12, which it was also proposed to exempt. More subtly, Curtiss also noted that much of the data had already been presented as a justification of the 1978 relaxations and should not therefore be used to justify a further round of relaxation.[15] In addition, Curtiss indicated that data were emerging suggesting that host strains survived better than was previously thought in hostile environments. The main direction of this argument was uncertainty, a view shared by interest groups opposed to such wide exemptions. Not least it was pointed out that such proposed revisions would come before risk assessment experiments being organised by the NIH itself were complete, and which were designed to test *E. coli* further. A final point was that the RAC meeting which endorsed the proposal to exempt the *E. coli* K-12 experiments, and other changes, took a vote when the majority of its members were absent. Consequently, the Director, NIH did not exempt *E. coli* K-12 work, but did accept recommendations on containment levels which had also been proposed.[16]

From this US experience of risk assessment, in the context of guidelines revision, it is apparent that it is difficult to be convincingly rational when uncertainty is commonplace. Much of the emphasis in the relaxation of guidelines related to lack of evidence 'to the contrary'. That is, it was argued by many that work with recombinant DNA techniques over the years had not revealed any hazards. However, specific attempts to identify hazards were a very small proportion of the total work, thus suggesting that inductive logic was important in altering perceptions of risk. Because something had not happened, the view strengthened that risks using *E. coli* were not likely to manifest themselves, a questionable process of logic. Some scientists also began to argue that because recombinant activity could occur in nature, with exchanges of genetic information between species involved,[17] then some earlier fears were unjustified. Even if genetic information can be thus transferred, it is not a valid argument to say that because a process occurs in nature it therefore does not require regulation when duplicated by man. Nature has demonstrated time and again that it can be dangerous to mankind through viruses and pathogenic organisms. Consider the effort to fight nature in eradicating smallpox, and more recently in tackling the AIDS virus. Until the natural mechanisms leading to new viruses, for example, are fully understood, we cannot assume that genetic transfer is harmless. Not least the advent of thousands of laboratories around the world undertaking the manipulation of DNA must add significantly to the total rate of successful genetic transfer. Nevertheless the actual use of genetic manipulation has revealed more fundamental knowledge of how genetic processes operate, including the greater complexity of expression mechanisms in eukaryotic, or higher, organisms, which in itself has suggested that accidental expression of eukaryotic DNA in prokaryotic hosts is unlikely. It is really this latter type of information that has legitimised the relaxation of conjectured fears. Thus the relaxation exercises owe much to new conceptual models of genetic processes which not surprisingly are more developed than the limited models of these processes which underlay the initial calls for caution. With the more limited model, the greater was the uncertainty regarding the hazards. From the beginning the dilemma was that only the application of the new research tools could really lead to theoretical knowledge of fundamental genetic processes and *understanding* of any risks.

Risks were legitimately conjectured given the absence of certain knowledge, or, put another way, there were holes waiting to be filled in the conceptual model. In that uncertain situation caution was commendable.

The heartsearching within the US over guideline revision was followed closely at the international level, while, in the meantime, GMAG for its part was embracing a quite different type of revision. Adopting the American way of publishing the proposed revisions for public comment, the proposals first appeared in *Nature*, in November 1978.[18] GMAG's new assessment method was based on a categorisation first outlined by Sydney Brenner, and was firmly rooted in traditional 'event tree' analysis. This is an approach to assessing overall risk in an activity where possible events at any point, in a complex set of events and alternatives, are allocated a probability of occurrence and the probabilities are then totalled for all possible routes to disaster. The sequences would follow from some initiating event. In diagrammatic form the analysis would look something like a family tree, but instead of offspring, at each point in a sequence of events possible alternatives would be indicated along with their probabilities.

Initially introduced for a trial run in parallel with the existing procedures, Brenner's alternative assumed a break with the foundations of both the earlier Williams guidelines and those of the NIH. He suggested that the possible pathways by which a manipulated organism could penetrate a human host, gain access to susceptible tissues and express products should be estimated. Thus in order to begin assessment of individual proposed experiments, additional information would be sought from experimenters. Proposals would have to include probability estimates of components of risk given the following labels: *access factor*, or the probability of escaped manipulated organisms entering the body; *expression factor*, or the likelihood of expression of the foreign gene into protein; *damage factor*, or the probability of physiological damage in the recipient's body.[19] GMAG recognised the difficulty of giving precise figures, but it argued that it would nevertheless like the local safety committee to provide 'approximate orders of magnitude'. A GMAG technical panel would make further

assessment before GMAG allocated containment requirements. Because many researchers might not be interested in
trying to achieve expression (merely wanting to obtain
quantities of genetic material for analysis) then they would
find the containment downgraded from the earlier guidelines,
but industry, which invariably would want to try for
expression, would find themselves relatively penalised. *Nature*
produced an interesting critique of the new procedures which
recognised the obvious difficulties of making probability
estimates but noted that although tests of individuals'
allocations of risk differed markedly, their rankings were
remarkably consistent.[20]

Despite the adoption of the new GMAG method of
allocating containment categories, it was the difficult
assignment of probability which hindered the application of
traditional risk assessment approaches. GMAG's allocation of
containment could function on the reasonably consistent
relative rankings of individual experiments, with the promise
of being very flexible as new knowledge became available.
With the benefits of hindsight and knowledge of the late 1970s
others have attempted to show that overall risks for individual
experiments are very low or have attempted to use classical
risk analysis as a basis to design containment facilities.[21] These
efforts again show the tendency to venture towards the use of
the familiar tools of rational logic and in no way detract from
the arguments that information of some comprehensiveness is
needed in such efforts or that the use of an empirical base is
not the only means to postulate risk. The initial institutional
responses which led to guidelines in a number of countries
were legitimately founded upon the state of knowledge of the
time. It was always the case that those who called for caution
also called for investigation of the risk. Krimsky, however, has
revealed the dangers of making too much of a little
information. An unfortunate consequence of this is the
growing trend amongst the scientists using the techniques to
suggest that the 'burden of proof' now lay with those who
would continue regulation. 'Innocent until proved guilty' is a
suitable premise for criminal law, but it is not an appropriate
analogy for the assessment of technological risk. For new
drugs to be acceptable, for example, clinical trials must

establish 'innocence' in advance. Similarly, motor cars in many countries face regular tests of their roadworthiness on the premise of accident prevention. It is, however, true that many accidents and disasters which have occurred in the past in a wide variety of areas could have been made less likely if more conjecture of possibilities had been applied, or known information more readily disseminated.[22] Perhaps the predominant feature of the case of genetic manipulation was that the regulations were being applied, not at the level of industrial products, but at the very level of inquiry at a new frontier in basic scientific knowledge. Again this reinforces the uncertainty argument. With time the question of risk has modified, but this is really a question of appropriate controls, and in turn should not influence the assessment of new risks which might arise as genetically engineered organisms are used in large scale industrial processes or are themselves deliberately released into the environment. It is a fact of life that it is unlikely that in the long term genetic manipulation will avoid public scrutiny given likely *increases* in the ability to modify all forms of life including human life.

Risk-Benefit Assessment
By the 1980s benefits of genetic manipulation had already accrued in a wide variety of areas, including quite simply the vast increase in fundamental knowledge. Drawn within the gamut of biotechnology, the techniques of genetic manipulation represent a functional tool in the design of both new products and production processes, often in combination with other contributing techniques. But in the early years of the genetic manipulation debate the benefits, like the risks, were also conjectural. In the quest for appropriate policies towards genetic manipulation it could be said that outside the realms of rationality there was indeed a politicised process of assessing risks against benefits. However, as with many issues of technological safety, there was little centralised risk-benefit assessment, although the Ashby Report in the UK and the NIH Environmental Impact Statement for the original US guidelines made some tentative conjectures. Generally, the few attempts made in the 1970s were little more than lists of speculative risks and benefits, with conclusions to the effect

that the benefits were likely to be very great and, therefore, the work should continue. This is meant less as a criticism than as an observation that at the time comprehensive risk-benefit assessment would have been enormously complicated, given the limits of knowledge. In effect, the whole transnational debate about genetic manipulation was a social and political airing of risk-benefit issues without the constraints of rational logic. Consequently, there were great differences of opinion amongst the transnational actors involved.

Scientists, as a group, invariably saw a future of increasing benefit, set against a future of declining risk – a powerful argument for the work to continue. But the decisions about safety were not to be taken by the scientists in isolation. Other actors were more uncertain in their beliefs, and saw broad participation in processes of assessing risks and benefits as critical. In this sense much of the politics was about the search for legitimate decision systems while superimposed on all of this was the nature of inter-system interaction. A key factor here was the response in one decision system if another changed its assessments or procedures. It became quite apparent that scientists and industry closely monitored the international process of guideline relaxation and moves towards legislation as they affected individual states. Indeed, threats of moving abroad to use lower containment levels for specific work were at times used as political ammunition. A Swedish firm, for example, in 1979 threatened to go abroad if permission to use recombinant DNA techniques was not granted.[23] Earlier in 1978, a US scientist, Stanley Falkow, returning from an international symposium in Milan, wrote to the Director, NIH in response to a request from the latter for him to consult colleagues on risk assessment evidence. However, he also returned his comments on relative guidelines:

Aside from these positive aspects of my travel, I confess that, in another vein I found the meetings most distressing. It is painfully obvious that because of the very restrictive nature of the NIH guidelines as well as the bureaucratic wall that the guidelines have spawned, American biologists can no longer expect to keep pace with either Western European or East European science.[24]

Falkow was commenting at a time when the NIH guidelines

were more stringent than, for example, the UK guidelines. Sir John Kendrew made the following point after the US revisions of 1978, and in the context of the Wye College conference of 1979; 'When we had the conference at Wye in Kent with the international group of scientists involved last April, many of the Americans came over having got their revised guidelines and really trying almost to steam roller us into acceptance that all the regulations should be swept away.'[25] Others, however, welcomed the NIH revisions as an opportunity to press for UK guidelines reductions. An influential report on biotechnology has stated: 'We recommend that GMAG considers urgently the possible prejudicial consequences to British industry if controls on genetic manipulation in the United Kingdom are more severe and restrictive than in other countries.'[26] The same report noted that although GMAG's flexible approach made UK controls less restrictive than the NIH guidelines for some work, overall they were more restrictive. This illustrates an important point. Pressure for relaxation of guidelines did not only operate on the basis of comparing guideline packages, but often identified particular experiments or types of experiments treated more severely in the home guidelines.

Thus internationally there operated a 'least common denominator' effect, whereby at revision times there was some urgency towards selecting the level lowest in a survey of national guidelines. In addition to this, industry and scientists were well aware of growing competition in the promotion of biotechnological research. European states in particular were often host to comparisons between the rates of biotechnology development internationally. Tied to observations of the relative position of domestic research were requests for national support in promoting the technologies involved. In turn, these requests for support influenced the calls for guide-line relaxation. It was, in fact, this growing sense of competitiveness which made harmonisation of guidelines within Europe increasingly difficult, especially if legislation was to be involved. Although harmonisation of guidelines, supported in principle by international organisations like the ESF, EMBO and COGENE, might have provided the standard within which competition could have flourished, there was one predominant flaw. Unless *all* states were within such a standard there would always be the 'free-rider' problem. One state lowering guidelines significantly below those of other states might attract industrial research, perhaps from the multinationals, or even the brainpower of individual scientists who would move abroad. Harmonisation in such instances would itself become a constraint in the eyes of some. Flexibility in a rapidly changing situation became more desirable. By 1980 the general feeling of a COGENE meeting was that the very development of international guidelines might only serve 'to preserve guidelines'.[27]

In the last resort, as with any new technology, it will be society at large which will determine the final utility of biotechnology. Scientific, economic, social and political forces will all be at work in determining the long-term success of the likes of genetic manipulation. But one feature is perhaps most political. There is no doubt that genetic manipulation will have its uses to any group or state wishing to manufacture biological weapons. We may also find future temptations to interfere with the genomes of human beings, despite efforts at controls in this area. Roy Curtiss, a key figure in the discussion of potential risks, and also a key figure in the designing of safer hosts based on *E. coli*, has noted that how our new knowledge is used may involve choices outside the scientific community:

I have spent two and a half years of my scientific life in taking the cautious approach to provide safer systems for cloning and to establish to my complete satisfaction, at least, that no harm will come from this research unless it is a conscious decision of society to use the knowledge gained from recombinant DNA research for purposes over which the scientific community has no control.[28]

Social accountability has been demanded by various groups around the world, whether Friends of the Earth, the British Society for Social Responsibility in Science, the critical media or even noted individuals. Many scientists have been amongst their numbers. With the eminent scientists who originally called for caution was it really surprising that less informed or knowledgeable people might be even more in fear? But one group active in demanding social accountability came to worry the scientists more than any other; the legislatures.

QUESTIONS OF CONTROL

Legislation

Not the only country to consider the use of legislation in order to extend coverage of their guidelines to all users of recombinant DNA techniques, the exercise was nevertheless particularly intense in the United States. However, scientists and industrialists in many other states also kept a keen eye on events in Washington. What happened in the US could happen elsewhere. Much of the literature commenting on genetic manipulation emphasises the experience of the United States and delves deeply into the issue of legislation.[29] Taking an international perspective, the case of the United States should not be overemphasised, other than in the context of its altering perceptions seen elsewhere, especially as some countries such as the UK, Sweden and Switzerland had existing statutes of considerable assistance in guideline implementation.

We could also ask whether legislation need have been a problem in the United States itself. Sometimes seen as an attack on 'freedom of inquiry', US legislative proposals were originally not at all a challenge in this direction. For Burke Zimmerman, of the Committee on Interstate and Foreign Commerce, legislation was not the response to the perceived hazards – that was in fact the NIH guidelines themselves – but was the response to exemption of some researchers; namely those who were not NIH-funded.[30] Such a challenge might not have been the intention, but it certainly could have been the case if some of the legislative drafts had ever become law. We have already seen how the Federal Interagency Committee was reponsible for the move towards legislation, eventually to be simultaneously introduced into the two Houses by Rogers and Kennedy. Kennedy's interest had also extended to writing (along with Senator J.K. Javits) to President Carter to request executive action over genetic manipulation. In effect the Bill was the result of the Interagency discussions and represent-ation from the President's office. However, the problem was that by this time (April 1977) a number of other Bills were being introduced. Whereas the Kennedy–Rogers Bill was initially a sober attempt to extend the guidelines to all sectors of the research community and to the uses of the products of genetic manipulation, some of the other Bills worried the researchers. Such was the concern that many scientists offered support or at least acceptance of the proposals in the government Bill.[31]

New Bills were not the only problem. Amendments to the Rogers Bill, for example, took the penalty for violations from $5000 a day to $50,000 a day.[32] Yet in the comparisons between all the legislative proposals, one issue came to the fore. This was the question of whether or not local government could go beyond the requirements of Federal legislation and impose harsher restrictions. Local and vociferous debates, for example the Cambridge altercation, which were often given extensive coverage in the United States and elsewhere, had already given the scientists cause for concern. Less noise and more rationality were seen as preferable by most scientists rather than many local confrontations with 'ill-informed' pressure groups. Consequently, scientists turned from offering limited support for legal extensions of NIH guideline coverage to becoming active lobbyists against restrictive penalties and local regulations. After much lobbying and with the withdrawal of Senator Kennedy's support for the government Bill, legislation was left to wither in likeliness and recede completely in its original form. US industry was also left with only voluntary compliance.

The success of the organised lobbying by the scientists and groups like the American Society for Microbiology (ASM) and the Harvard-based 'Friends of DNA'[33] was intrinsically linked with the emergence of the risk assessment activities purporting to show a decline in estimated risks. Gorbach's letter written immediately following the Falmouth workshop was very influential, despite some attempts by a number of participants at the workshop to show that Gorbach both overemphasised the degree of consensus and extended it to some areas where there was insufficient evidence for a consensus to emerge. In addition, results from the laboratory of the cautious Roy Curtiss led him to a change of heart, and to publish both his results and the process of his conversion.[34] These views greatly impressed themselves on the legislative process, with significant supportive lobbying.

Much of the problem with the process of US legislative

activity centred on the range of differences in the dozen or so proposed bills. Kennedy's Bill became associated with establishing a nuclear energy-style commission to oversee all work, while Rogers developed his Bill to place regulatory authority under the Department of Health, Education and Welfare (HEW). The Rogers Bill was favoured by scientists until an amendment made local extensions more likely. Other Bills proposed, for example, only a few authorised research institutions, investigative commissions of inquiry and higher fines. With the environmentalist lobby pushing for local controls, and thus further opportunities for public participation, the whole issue of US legislation was very controversial. With the success of a now politicised and more adept scientific community, one final approach to legislation emerged. Zimmerman tried to draft a Bill which the main institutions would support. It extended the coverage of the NIH guidelines, placed regulation under the HEW and was designed to be a two-year interim piece of legislation. Harley Staggers introduced it but it died of lack of interest. The ASM and the Director, NIH among others supported it to a degree, but the environmentalists opposed it, hoping for something stronger in the next session. Norton Zinder has summarised his view of the attitudes of the scientists in saying:

I believe the scientific community would have accepted the simple codification of the NIH guidelines into the law of the land. These bills were far from that. They set up vast bureaucracies, cumbersome licensing, harsh penalties and tedious reporting procedures.[35]

Despite earlier reticence about guidelines, US industry fully endorsed the idea of Federal legislation. The principle of extending cover of the NIH guidelines to all users of recombinant DNA techniques was quite compatible with their promise to comply with them on a voluntary basis anyway.[36] Most of the debate surrounding US legislation occurred before industry became involved beyond the laboratory level, where industry was in any case well used to safe practice. It made sense to comply voluntarily in that any future realised hazard could be legally defended on the basis of adherence. The vociferous lobbying was thus on the part of the scientific community in general, with industry maintaining a lower profile in order to avoid environmentalists further questioning their motives.

Consequently, in a climate of general uncertainty, and with misperceptions at times of what some Bills proposed, an unnecessarily complex process of political conflicts led to a lack of action. However, 'no action' was by 1979 considered a significant victory by the scientist activists opposed to harsh legislation. Those pressing for strong legislation gained early momentum from spectacular press coverage promoting unfounded levels of fear. Challenged by the language of scientific theories and empirical evidence, supported by equally strong lobbying, they eventually lost out.[37]

It has already been argued that the degree of international borrowing of the US and UK 'packages' was extensive. Not only were the guidelines of these two states adopted, but their attitudes to statutory support for their guidelines were well observed. Thus West Germany attempted to develop new legislation while others considered the adaptation of existing statutes, much like the UK. But in Europe the issue of legislation was brought fully to the international level through the proposed European Community Directive. Throughout all these cases the challenges to the idea of legislation were similar. It seems that one of the main difficulties with all approaches to legislation from the local city level through the national level to the international regional level, as attempted, was the inability to present an appearance of sufficient flexibility. Whatever one thought of the risks involved, it was apparent that if allowed to continue, the work would progress rapidly. Thus, attempts to mould a more slowly moving bureaucratic response, while keeping up to date, and in a generally controversial area, were full of difficulties. From this, in summary, a number of points are important.

The inflexibility of the bureaucratic process supports the argument that the initial actions of the concerned scientists were creditable and fast, particularly in that they took their concern and caution to a fully international level from an informal stance. The problem of more formal controls has, of course, led to the varieties of approach discussed in this book, and at the international level standardisation has been elusive. This problem has had two stages to it: first, there has been the

failure of effective harmonisation of guideline recommendations, to the extent that the 'lowest common denominator'
effect has characterised revision exercises; secondly, with such
differences and, indeed, international competition towards
relaxation of guidelines, international regional legislation
became a non-starter.

Monitoring and Sanctions

Had legislation been more widely adopted internationally,
then provisions for monitoring and enforcement of guideline
adherence would have been more stringent. In particular, the
punishment for infringement through carelessness or wilful-
ness might have taken on a more explicit use of the concept of
deterrence. However, in the context of a general failure of
legislative proposals, it is necessary to provide some analysis of
the actual nature of ensuring compliance with the appropriate
guidelines among genetic engineers.

An initial assumption made in most cases of establishing
safeguards was that scientists involved would not deliberately
violate recommended containment on specific experiments.
Many saw the recommendations literally as 'guidelines' of
most appropriate containment, particularly when initial
derivatives of the Asilomar II suggestions were applied.
Caution in the early days characterised the scientists' approach
to the issues, making the need for agreed precautions dominate
the agendas of discussion rather than means of enforcement.
Critical in understanding this perspective is the central
importance in science of peer review. Openness in publishing
results and sharing the knowledge acquired from research
would make secret violations of limited utility to the majority
of researchers. Knowledge of the facilities in different
laboratories could quite easily be related to the requirements
for different types of experiment. Indeed, the greater the
potential violation, the more likely this would be observed
when results were shared. It was this principle which in retrospect confirmed international adherence to the moratorium
following the Berg letter.

As long as there was consensus on the nature of the risks
and the need for guidelines both for safety and to allay public
fears, then perhaps the above approach could suffice. Given

disagreements between sections of the scientific community and the pressures to achieve success in career terms, then the strains on such monitoring might show. Further, in research centres with commercial interests secrecy might prevail, with limited publication of results and thus limited peer scrutiny. As the industrial application of genetic manipulation expands, shifting the balance of work away from academic centres of research, then peer review will become less relevant as a means of ensuring guideline compliance.

In the event violations did occur. In September 1977, Nicholas Wade reported an early breach of the NIH guidelines by researchers at the University of California, San Francisco. The research group broke the guidelines by using a vector before it had been certified by the Director, NIH, although after the RAC had recommended its acceptance. The point is less the seriousness of the violation, than the fact that some competitive advantage accrued to the team given the high interest at the time of genetically engineering human insulin.[38] In 1980 a further violation arose when a researcher used the wrong virus, from two sent to him some years earlier, in an experiment at the University of California, San Diego. The violation was spotted by two students in the laboratory involved, refuted by the researcher, Samuel Kennedy, and subsequently confirmed by an independent review, which concluded that the use of the particular virus was either an error or deliberate.[39] Little action was taken against these 'offenders', although in the second case the NIH on examining it decided that as Kennedy had resigned his post and was no longer funded by the NIH there was nothing further they could do anyway and let the matter rest.

The *cause célèbre*, however, of the violations was truly international in its ramifications, and involved Martin Cline, this time of the University of California, Los Angeles. The case illustrates complications concerning the lack of international harmonisation of guidelines and moreover the possibility of going abroad to do work explicitly banned at home. Cline used genetic-manipulation techniques to alter bone marrow removed from two patients, which after alteration was replaced. He was hoping to treat a gene-deficient disease known as thalassaemia which interferes with the production of haemoglobin, causing anaemia. However, under NIH rules all experiments involving human DNA were banned, and Cline failed to get approval for such work. In response, Cline, who was NIH-funded, undertook the experiment abroad. One 21-year-old girl was treated in Israel and another girl of 16 was treated four days later, in Italy. It is clear that at least in the Israeli case the authorities were misled, believing that Cline intended to use unaltered bone marrow cells. Indeed, the authorities in the Israeli hospital took great care to make sure that Cline had no intention of using genetic manipulation, and were consequently disconcerted. In both cases the girls were neither significantly improved nor harmed by the treatment. Cline had seen his work as simply the next stage in progress from research with animals and took a unilateral decision to make that advance abroad.[40]

Cline, chief of the division of haemotology–oncology, resigned his position like Kennedy, but remained as Professor of Medical Oncology. After investigation of the case, the NIH response to the violations was to require all future recombinant DNA work by Cline to need special permission, and that review committees assessing his future proposals be furnished with details of his contraventions. They also recommended that the three institutes at that time funding Cline to a total of $600,000 might decide whether any of his current research money should be withdrawn.

With strong competition in science for both recognition and funding, peer censure is quite a strong deterrent within the limitations of academic tradition, especially where revealed transgressions are the subject of considerable publicity. Had the violations been more serious, or in Cline's case some years earlier, then the responses would most likely have been much greater. Despite this, there is some indication that the NIH was reluctant to adopt a punitive role. Legislation could have overturned this situation both in the US and elsewhere. In the event most states relied on peer pressures and the threat to future funding as the means of ensuring guideline effectiveness. It could be argued, however, that as scientists perceived risks to decline, then they would perhaps be more likely to ignore guidelines if they saw them as too stringent. Curtiss said as much with regard to impending US legislation if it were to prove too restrictive.[41] On the other hand, the UK Health and Safety at Work Act, although legally providing facilities for censure, might prove difficult in application as the Birmingham smallpox case revealed in a similar situation, where action in the courts against the university was unsuccessful.

Monitoring of safety adherence in most countries was usually the responsibility of a 'local' safety committee, reliant on peer pressure and administrative direction, and a national committee of some sort which could control or recommend sanctions where they existed and were applicable. As far as scientists in the public sector were concerned, compliance was probably sufficiently under control. For industry, the acceptance of guidelines was ostensibly voluntary, but with the firm's reputation sensitive to backlashes against any potential infringements. Internal scrutiny would thus surely exist, but the results of any 'inquiries' into non-compliance with guidelines would probably remain veiled. As most states developed no specific controls over industrial activity in the laboratory, the final safeguards were effectively their existing statutes for dealing with pollution and the possibility of private legal actions for compensation in the event of anything untoward happening. Nevertheless, it has been an international weakness that has treated the public and private research laboratories differently. However, the Cline case also raises the serious question of national differences in guidelines. What was there to stop a researcher or a company moving abroad if they felt the home restrictions could be avoided? Cline in the event faced the scrutiny of the NIH when he returned, and the threat to his future funding. But industry would not face such restrictions, nor necessarily would other nations' scientists. Whatever the difficulties of harmonising guidelines — and these have been addressed — there are also consequences of harmonisation not existing.

The situation is, however, becoming more complicated as some genetic manipulation research is reaching the stage of taking modified organisms out of the laboratory and introducing them into the environment. New policies regarding safety and environmental monitoring must follow in many countries and perhaps with international coordination from one or more of the agencies we have discussed. In this sense the anti-legislation lobby had a point in raising the question of flexibility in guideline implementation and responsiveness to changes in knowledge. Yet against this is the ever-increasing sophistication of biotechnology leading to more complex overlaps between activities currently overseen in safety terms by numerous agencies. Laboratory research, transport of materials, scale-up of production processes, and the environmental impact of products may all need inclusion in some safety and monitoring framework. In addition the intermixing of the old and the familiar with the innovatory scientific advances such as genetic manipulation may bring various safety provisions into interaction on an increasing scale, perhaps leading to revisions in all areas concerned.[42]

SUMMARY

This chapter has highlighted some of the substantive issues that have arisen from the development of genetic manipulation techniques. Risk assessment, risk-benefit assessment, legislation and monitoring of safety provisions have all reflected politicised interactions between groups with different interests. These issues have not been confined within national borders but have been revealed in a transnational context where events within individual countries have influenced activities within others. After introducing some of the newer international developments within industrial biotechnology, in the next chapter, Chapter 8 will return to the formal transnational analytical framework.

7 Industrial Biotechnology in an International Environment

Coincidental with the decline in perceived hazard and the reluctance to move towards formal international harmonisation was the growth of industrial attention to the promise of genetic manipulation. By the mid-1980s certain features of this growth were most apparent. National governments, in many states, had demonstrated the importance of their role in promoting biotechnology, given the competition in international markets, while research and development services themselves have become sought commodities. Associated with these trends has been the overall question of finance, much of it sought competitively by individual firms or joint projects and often with difficult problems of establishing the commercial risks involved. In a technology which is so research-dependent, it is not surprising that knowledge protection and international patenting issues are also notable. Yet, perhaps more fundamentally, biotechnology in the not too distant future will have quite profound impact on both international trade patterns, with challenges to traditional supply sources for certain commodities, and the linked question of the relationships between the developed and developing worlds. Finally, large-scale production processes may themselves bring new safety questions.

We have already seen that the European Commission, in considering guideline harmonisation and developing collaborative research projects, had to account for national competitiveness in the promotion of biotechnology. The US, the UK, France, West Germany and Japan represent just some of the developed countries with provisions for 'national' support in the development and exploitation of biotechnology. This support is, however, notably different in the United States where government funding has not been of importance at the industrial level. American firms may lead the world in exploiting the newer biotechnology but they have been spawned from the competitive entrepreneurial ethic of that nation, directly reflected in the availability of private sector venture capital and assisted by a tax structure geared to help new firms. Over 400 firms have biotechnological interests, and half of these specialise in this technology. In addition, US predominance derives from their large multinational pharmaceutical and chemical companies, providing marketing expertise and further sources of funding, which initially proved to be sponsors of smaller specialist firms and now are developing their own in-house expertise. Also supportive is the generally healthy state of US research in basic biological and related science, which has enabled close cooperation between academics and industry. In sum, it is the very success of the US biotechnology groundswell which has provided the shot in the arm to stimulate national biotechnology promotion in other countries.

National strategies must, however, take account of a whole host of factors ranging from the traditional expertise evident within the country to the assessment of appropriate international markets amenable to exploitation. There is also the need to balance the commercially viable with the development of long-term technological skills within the country's industry. Many studies have been made to account for national requirements, often arguing the importance to the country concerned of not falling behind their international competitors.[1] The Spinks Report, for example, argued for far-reaching government attention to be paid to both research and development requirements in the UK. Research Councils, various government departments, industry and university funding were focused upon with an emphasis on harmonising efforts, even to the extent of suggesting that the government establish, through the then National Enterprise Board, a research-oriented biotechnology company. Such a company would bring public and private finance together. 'Celltech' resulted in the same year as the report. It was further advantaged in that an agreement was made between the Medical Research Council and Celltech that the latter would have the first option in exploiting scientific advances made in MRC laboratories. In its turn the MRC in 1981/82 accounted for £22 million of the research councils' total of £29 million expenditure on biotechnology.[2] Part of the agreement included a provision that scientists' publication of research might have to be held up while decisions were taken on patentability. In return the MRC would benefit by receiving a 'substantial proportion' of any royalties.[3] Deemed successful, the agreement by 1984 had been extended for a further five years with modifications which in part took account of Celltech's growth and experience but which also brought some restrictions on the exclusiveness of Celltech's rights to exploit MRC work.

Within the UK further promotional policies have since developed, centred mainly on the efforts of both the Science and Engineering Research Council (SERC) and the Department of Trade and Industry (DTI), but with support from other relevant bodies. As Roy Dietz of the DTI's Biotechnology Unit puts it: 'there is little taste in the UK for grand imposed strategies; the approach is pluralistic...'[4] Strategy as such must come from industry itself, based upon its accumulated experience. The DTI, for example, will only give grants for industrial research and development projects up to a maximum of one third of the cost, to encourage the application of commercial judgement, although it will assist in the supply of strategic advice. The result was the establishment of a £16 million, three-year programme of assistance in 1982. Following the Spinks Report the SERC established its fourth ever 'special directorate', the Biotechnology Directorate. This Directorate, which includes industrial representatives, oversees the research council's biotechnology sponsorship in basic research, training and industrial collaboration.[5] Genetic manipulation is one of the important components of these national policy agencies.

Yet the British approach has not been without criticism, usually of the form that even still not enough is being done by government to foster a climate conducive to transferring the country's internationally leading basic research to industrial applications. Some apprehension was expressed in a report by Dunnill and Rudd. In particular they drew attention to the choices to be made regarding the identification of products and services that should be targeted by British biotechnology. Warning of the danger, for example, of focusing on too narrow a range of obvious targets they argued:

it is essential for the UK to seek those niches in the terms of skills, local market strengths which will aid exporting, and the existence of appropriate companies, which will help it withstand fierce competition. It will also be important for government supported research to be bold and to pursue the less obvious goals.[6]

Their observations extended to the position of UK large firms that have investment resources and have not been very active in channelling funds into new venture companies actively trying to develop new production processes and products. In comparison large US firms have been important supporters of their domestic venture companies while Japanese and other European large firms have also funded US start-ups in lieu of a shortage of these firms in their own countries. With less access to the knowledge generated by venture companies, UK firms may find themselves at a competitive disadvantage.

In 1982 the French announced a marked stepping up of government support for biotechnology through support for both research institutions and industry. Impetus for this was linked in part to a perception that France lagged behind other industrial nations. Earlier, in 1980, a commercial organisation called 'Transgène' had been set up to exploit genetic manipulation with initial capital of 40 million French francs. Funding came from both the private sector and various French research institutions, including the Pasteur Institute, Strasbourg University and the Centre Nationale de la Recherche Scientifique (CNRS). Britain's Celltech had influenced the approach taken, and it was hoped that further successful firms would follow. Thus, as in other countries, the French were to see combined moves on behalf of industry, government and academics to exploit the promised fruits of the 1970's biotechnology revolution. Coordinating French strategy in developing biotechnology has been a team from the

Ministry of Research and Industry, supervised by a National Committee involving representatives of Ministries, Institutions, and industrial fields of biotechnology.[7]

West Germany finds itself with some of the world's largest pharmaceuticals, including Hoechst the largest, while the country's lack of venture capital has restricted the emergence of new biotechnology firms. Indeed, Hoechst and others such as Schering have turned to deals in the United States to gain biotechnology experience. Generally speaking, it is the established firms which are making the running in German industry. With the all too familiar expression of fears of falling behind international competition, government support for biotechnology and the promotion of links between industry and academics are seen as most important.[8] More recently the West German government has considered the possibility of developing joint development programmes with other countries, such as the UK, Japan and Sweden.

France, West Germany, the United Kingdom, the Netherlands, Switzerland and other European states are all conscious of the United States' lead in terms of its venture-capital funding for the exploitation of the new biotechnology and in the number of new firms it has spawned. But the attention of both Europe, with its increasing European Community cooperation, and the United States is converging on the potential Japanese growth in biotechnology.

Japan has already shown its ability to become a world leader in applying certain technologies and industries, often being particularly quick to take up ideas conceived elsewhere. Biotechnology shares some of these characteristics, especially as far as the new techniques of genetic manipulation and cell fusion are concerned, and Japan has now embarked upon acquiring skills in these areas with a vengeance. But in the case of biotechnology, Japan already is a world leader in the more traditional realms of fermentation and enzyme applications. There are around 200 firms in Japan researching and applying biotechnology knowledge, but in ways that have been described in one analysis as six times as efficient as their American counterparts. As Gregory puts it: 'The US's pharmaceutical industry has been spending about twice as much as the Japanese industry to develop only about one-third as many new products.'[9] This growth in effort has not been without obstacles. Japan has lagged behind the US in the development of a set of guidelines under which genetic manipulation research could proceed, in that it was not until 1979 that their own version became available. Additionally, Japan was acutely short of the scientists trained in genetic manipulation skills. Set against this, however, is the perception of a number of large companies that Japan, in its fermentation industry, possesses considerable expertise necessary to take the techniques of genetic manipulation from the laboratory to the large-scale industrial process, with unsurpassed quality control to boot.

Biotechnology's future raises some serious questions about the Japanese past approach to research and development where the emphasis has been on the adaptation of overseas technology. Today continuing research is critical in the development of new products, where lead times between the laboratory breakthrough and the marketing of a new product may be short indeed. One ministry, the Ministry of International Trade and Industry (MITI) sees its role as one of complementing the leadership of private enterprises by providing guidelines for the longer-term structural problem that cannot be dealt with by the market.[10] Part of their policy will be to encourage international as well as domestic cooperation in biotechnology, in reflection of the view that biotechnology in the long term will be vital in the progress of the world economy. MITI is active in establishing bodies in Japan to coordinate R&D through, for example, consortia and Research Associations. Attacking the infrastructure MITI supports moves to broaden university/industry links and to provide computer equipment and software to help overcome the scientific labour shortage. Additionally, within its own house, MITI is using its eight research institutes to research and disseminate basic scientific knowledge. Although other ministries are also engaged in the promotion of biotechnology there is a tendency for this to be in keeping with their parochial objectives.

If there is a role for government ministries it is in parallel with industry's own efforts, perhaps best assessed in terms of Japan's large multinational trading conglomerates, such as

Mitsubishi, Mitsui and Sumitomo, and its leading firms in the pharmaceutical, food and chemical sectors. Within the former, interdepartmental specialist teams find their activities spanning departments and reaching beyond the firm to other companies and research institutions both at home and abroad. These traders have even begun to invest and participate in foreign biotechnology companies; in US venture firms, for example. Nevertheless they have still shown themselves to be relatively passive in the taking of commercial risks in the newer areas of biotechnology preferring to channel their growing interests in biotechnology, according to Toshio Itoh, into their traditional areas of interest, such as sugar refining, oil milling and timber processing. At a minimum they are, therefore, important in bringing knowledge into Japan and in the longer term, experts in these firms may be attracted towards the establishing of Japan's own new biotechnology ventures, as happened in the past in the electronics industry.[11] Because of the lack of venture capital and new firms specialising in biotechnology, new products are predominantly coming from the established firms. In the longer term, the shortage of certain skills may, however, be overcome through the experience of traditional biotechnology firms. In the meantime a substantial number of new microbiological products have already been produced.[12]

Despite the worries of Japan's competitors its future impact in the biotechnology industry will not be without considerable changes within its economic infrastructure. However, Japan has adapted in the past on the grand scale and is likely to do so again. Competition is, therefore, already a national priority of most developed states as far as biotechnology is concerned, reflecting economic nationalism, and this contrasts to some extent with the apparent international focus of the dawn of recombinant DNA guideline development. But national efforts aside, some important aspects of the future development of the technology are likely to retain an international dimension. Finance is one of these.

Venture capital may have greatly assisted the growth of new biotechnology companies in the United States, but new firms in most other countries have been relatively starved of such funding. As a result the competition for funds, given the general reluctance of finance houses to support what they see as high-risk ventures, is quite transnational. It was the promise of bringing new products and production processes to world markets which encouraged the tremendous growth of new biotechnology firms in the late 1970s and the early 1980s. The problem has been that the time to progress from the research idea to a tested and sellable product was underestimated. Large financial outlays have been required to cover the research costs of these firms well in advance of any production returns on this investment. Consequently there has been some sorting out of the weaker firms and a number of takeovers as larger established firms have purchased biotechnology start-ups and thus acquired expertise. Funding for the more successful of the new firms has come from a variety of sources in addition to venture capitalists. Firms like Genex, Celltech, Cetus and Biogen have all, for example entered into agreements with other companies to undertake contract research and development, providing an essential service to larger firms such as chemical and pharmaceutical giants who may lack the in-house expertise or facilities. Finance has also come from going public with one or more share issues as the better companies have begun to demonstrate their worth. Limited partnerships have been seen as an attractive option for some, and notably for Genentech, the most successful of the new firms. Another common strategy has been to embark upon licensing agreements with larger firms to bring initial products to the market place through the production facilities and marketing experience of the more established partner. It is, however, a transnational pattern of interaction. Biogen has its corporate headquarters in the Netherlands and production facilities in both Geneva and Cambridge, Massachusetts. But like many of the new firms which are located in one country (predominantly the US) it has arrangements with companies around the world. There is a reciprocal transactional network involved. Established companies from around the world want access to the new biotechnology firms' expertise, while the new firms want both funding and increasing access to world markets. In a technology with such novel scientific techniques at its heart the whole area of financing is illustrative of another dimension to risk-benefit assessment. Finance for so much research and relatively few products is seen by all too many potential sources as high risk.

A second important international feature of the biotechnology revolution has been the protection of knowledge. This is particularly salient given that the search for knowledge is not just confined to new products, but is intrinsically tied up with new production processes applicable to old and new products alike. Patenting of inventions or new processes is at the heart of most research-oriented firms' activities, but in the realm of biotechnology there are particular characteristics. Not least is the fact that it took years in the United States to get the principle of patenting new forms of *life* accepted. The path-breaking claim, now quite famous, was granted to Ananda Chakrabarty in 1980, covering work completed eight years earlier. Using non-genetic manipulation methods, Chakrabarty produced an organism combining the characteristics of four original bacteria. Each of the original microbes contained plasmids capable of breaking down a component of oil. The new microbe could attack all four components, a feature which might have potential for treating oil spillages leaving harmless byproducts before the organism itself disappeared. To be useful in commercial terms a patent was vital as, once released into the environment, the bug would be freely available to anyone taking a sample. At the time of application the US Patents Office would not issue a patent on the product as new forms of life of this sort were not covered. However, it would grant a patent on the process of its manufacture. Following an appeal, the eventual judgment concluded that if all other criteria were met, then the fact that the product was alive should not hinder a patent, and it was duly granted.

The importance of patents in biotechnology has been powerfully demonstated by the application by two universities, of all bodies, for patent protection on the very techniques of genetic manipulation. The universities of Stanford and California, representing Stanley Cohen and Herbert Boyer respectively, were eventually awarded patents on both the techniques of genetic manipulation and, more significantly, the products of their use. The less controversial techniques patent was awarded in 1982 while the potentially more lucrative products patent took a decade of investigation before eventual acceptance in 1984. A careful strategy of licensing, where the charge is not set too high so as to encourage legal challenge, ensures both universities of considerable earnings. With the floodgates opened, the number of worldwide patents filed in biotechnology has risen from a few in 1979 to over 1300 in 1984 alone.[13]

At the international level, the issue of patenting was recognised quite early in the development of the new biotechnology. In April 1977 the World Intellectual Property Organisation (WIPO) convened a meeting in Budapest on the important question of the deposit of samples of new microorganisms. New 'inventions' for the purpose of patenting need only be described in detail and filed with the appropriate patenting agency in most countries. However, for microorganisms, the disclosure requirements in an increasing number of states appeared to necessitate the deposit of a sample of that organism in a special institution, costly to maintain, from which further samples might be withdrawn. The Budapest meeting was for the purpose of drawing up an international treaty to simplify matters. It was proposed to avoid the costly duplication of samples in every country where a patent was applied for, by allowing one deposit in a single state to cover subsequent patent applications in other countries. This would necessitate the recognition of 'international depository authorities' for storage and international provision of samples. The original proposal that WIPO should examine the problem came from the UK, 29 states attended the meeting, and the resulting treaty came into force in the UK, for example, in December 1980.[14]

Although patenting of microbiological products and production processes has become internationally acceptable, there are still a number of considerations that potential applicants need to address. In the first instance it may not be the case that filing a patent is the best course to follow. It may be better to opt for secrecy, rather than the disclosures necessary in a patent application, where the product may only have a relatively short market impact before being superseded, or where the patent may take a long time to be approved and the necessary disclosures assist competitors in the meantime.

Many new biotechnology firms have fostered links with academic research institutions and universities, and this brings together two different traditions regarding openness of information. Academics in keeping with the norm of peer review find themselves under great pressure to publish their work. Industry tends more towards secrecy to maximise the commercial gains from new products and production processes. Consequently, deals between these two groups may require the academic to suspend publication until after patent protection is awarded. In the US, however, prior disclosure can be made up to a year before the filing of a patent, which may help in this context. But in other countries any disclosure can invalidate patentability. Therefore, as the trend for academic/industry cooperation increases, in part because universities are finding it harder to maintain traditional funding levels in the face of cutbacks by government, these new constraints are being accepted.[15]

It is worth briefly looking beyond the immediate future to see what impact there might be on international trading patterns. Certainly the pharmaceutical industry, the chemical industry, healthcare, agriculture and the energy industry are all going to be increasingly affected. It is most likely that changes in one or more of these industries will influence important international trade patterns. Important as these changes might be within the developed world, they may have enormous significance in the context of trade between developed, semi-developed and developing countries.[16] This will be particularly the case where biotechnology is applied in order to create alternative sources of foods, chemicals and energy provision which offsets the dominance of any group of traditional suppliers. Developing countries, with their more limited economic infrastructures often find themselves dependent on a narrow range of exports with which to earn foreign currency and finance growth. In 1982 a biotechnology report for the OECD drew attention to some particular problems faced by developing countries.[17] Identifying agriculture as particularly significant in this respect, the OECD report noted that the industrialised countries are heading towards self-sufficiency in certain commodities, as a result of high technology, while at the same time they are attempting to increase the export of biotechnology to the Third World. Developing countries may, therefore, find themselves losing markets for the export of certain foodstuffs while experiencing pressure to move into yet another high-technology area. An important example of changes to a traditional commodity has been the use of biotechnology to produce alternative sources of sweeteners, to the extent of almost causing a collapse in the sugar cane market. Since 1975 high-fructose corn syrup (HFCS) has dramatically challenged the previous monopoly of cane and beet sugar, and by 1983 HFCS accounted for more than 30 per cent of the US and Japanese sweetener market.

Technical dependence may be the result for many developing countries as they strive to acquire skills in biotechnology, but for some semi-developed or newly industrialised countries biotechnology is already well underway. Alcohol and biogas production represent the sort of technology where the leadership is held by countries like Brazil, China, Mexico, India and the Republic of Korea. And as Edgar DaSilva of UNESCO notes: 'Other factors instrumental to economic development are a balanced approach to the problems of technology transfer, the development of trained manpower and the judicious use of natural renewable resources.'[18] Consequently, most developing countries feature biotechnology in their national development plants. This momentum for acquiring biotechnology skills is reinforced by the need of many developing countries to improve food production efficiency to cope with ever-increasing populations. Technology transfer may, therefore, become an even more important issue in biotechnology than in other technologies. However, great care will be necessary to ensure that biotechnology skills and processes are selected and nurtured in these countries in harmony with their existing economic infrastructures, and not simply for the sake of technology acquisition. Combatting the threat of HFCS may be best achieved by finding new uses and markets for traditional sugar crops and their potential byproducts than by replacing those products with new ones.[19]

One developing country which has taken steps to ensure the acquisition of biotechnology skills, in the face of some adversity, is India. In 1981 the United Nations Industrial

Development Organisation (UNIDO) put forward the idea of an international centre which would encourage the transfer of genetic manipulation skills to the Third World. India put forward a case for the centre to be located in New Delhi. The government offered to donate $20 million for initial fixed assets followed by $1 million for at least five years to assist in the centre's funding. By 1982 India had also embarked upon a national programme in biotechnology funded to the level of $40 million over three years. UNIDO, however, opted for a controversial proposal. Its committee argued that New Delhi lacked advanced research to act as an infrastructure for the UNIDO centre and was several hours' flight-time from other centres which did possess microbiology experts. Pakistan and Cuba were similarly rejected. Controversially, they proposed to site the centre in Belgium, Thailand or Italy. To many the inclusion of two developed countries in the list went against the idea of fostering Third World skills. India continued its efforts and in January 1984 the UNIDO preparatory committee reached a compromise recommendation that the centre be in fact split between New Delhi and Trieste. The expertise argument was rejected in favour of the amount of funding the Indian government was offering. Trieste would specialise in industrial microbiology and energy research while New Delhi would focus on agriculture and human and animal health. Training for Third World researchers was also to be offered.[20] The Indian case displays a combination of keenness to gain new biotechnology experience and the requirement of some existing infrastructure and expertise to host the newer technology. Technological transfer always faces such constraints where skills have to be acquired by, and then spread within, the developing country. That country may also have to contend with a future 'brain drain' as the best new technologists move abroad where potential earnings and research facilities are better. In a technology like biotechnology, where the global impact itself is new, this is likely to be even more the case.

The impact of biotechnology on international trade patterns will be noticed in developed and underdeveloped countries alike. With much of the world's technological expertise residing in the big national and multinational corporations, then competition and the market success of each of these large firms will influence the national balance of payments and trade strength of their home countries. Where biotechnology leads to new sources of products traditionally provided by particular companies then economic muscle may be transferred. The sweetener case illustrates the extent of change that can happen. Protein production may be another long-term example if single-cell proteins make more impact. Human dietary customs are such that our tendency to eat meat necessitates the transformation of more than half of the world's plant-produced protein into animal protein at an average yield of only 12 per cent. Applying microbiological knowledge it has become possible to produce single-cell proteins which can probably thrive on a variety of raw materials such as paraffins, methanol, wastes and cereal starch.[21] A major contribution to the world's animal feed could potentially develop, even though production at the moment, led by ICI, is quite low. Competition with soya bean produce would ensue. Perhaps the more significant areas of change will be noticed where chemical and agricultural production overlaps. Chemicals produced from biomass may eventually provide a considerable alternative to chemicals derived from non-renewable fossil resources. Progress in key areas of biotechnology will be needed if a significant shift in emphasis is to be realised. Many chemicals will never be produced by biological methods while most complex biological macromolecules will remain beyond the skills of chemists to synthesise. But one type of chemical is wide open to biotechnology and chemical industry: the oxychemicals such as organic acids, amino acids, motor fuels and gases. In the long term, biotechnology must increase in its international impact across a range of areas such as the above. It is for this reason that the competition is so keen between the international and national government sponsorship programmes, the traditional biotechnology industries, the new start-up firms which hope one day to be large firms, and the non-biological industries which may face new challenges. In keeping with the interdisciplinary science involved and the mixture of old and new production processes, it may be the transnational conglomerates who will prevail in the long run with their ability to tap into a range of expertise. Dependencywise, this could continue to weaken the position of the underdeveloped world.

The global spread of the use of genetic manipulation and associated techniques will also bring new safety issues. These will appear at two levels: first, basic safety practices will be required in all laboratories, and especially those undertaking higher risk experiments, whatever the country and its level of development; secondly, at the industrial application level all countries need to consider the safety requirements imposed on both their own domestic firms and 'experienced' multinationals. Despite all the problems associated with the achievement of international guideline harmonisation for safe laboratory practice this is probably easier than achieving a similar harmony in industrial biotechnology. At the laboratory level there is always the possibility of borrowing 'off-the-shelf' guideline packages, such as those of the US. Whereas at the industrial level, even in the developed countries which lead the field, there are by and large no clear parameters for the safety of bioreactors and environmental release. 'Off-the-shelf' borrowing would be even less appropriate for the developing countries. Complex national technological infrastructures tend to foster, in turn, complex bureaucratic responses on safety issues. The case of the United States shows us the problems of agency jurisdiction, for example, in the regulation of industrial application of genetic manipulation. Existing statutes have to be assessed for overlap with any potential new statutes. How, then, can the developing countries hope to cope with regulating the activities of multinational firms in their urgency to acquire biotechnology expertise? If the lowest common denominator effect expands beyond the laboratory guidelines, then pressures may build up as companies exploit national differences in the stringency of safety requirements. Third World countries may be disadvantaged in this respect, or may try to strengthen their bargaining position by exploiting lower safety provisions to entice companies in. A corollary of this might be for developed countries, in competition with each other, to use specific safety rules as non-tariff barriers.[22]

SUMMARY

Whatever promise biotechnology has for the future, we shall not be able to ignore the issues all too briefly introduced in this chapter. As competition increases and the new biotechnology carves out its industrial niche, it will become another significant industry tied into the complex economics and politics of international trade and development. Safety policy may remain essentially nationally based, but any potential hazards will surely be transnational. As the biotechnology revolution leaves the laboratory on an ever-increasing scale many further questions will deserve our attention.

8 The Transnational Perspective

The issues raised by the biotechnology revolution which began in the early 1970s have undoubtedly left their legacy. Debates and assessments, policy-making and recriminations, and outright politics have involved a whole host of individuals and organisations, within, across and between states. Some analytical order may now be derived from a formal return to our transnational framework.

The publication of guidelines, the creation of provisions for their implementation, and the consideration of legislation all reflect the activities of decision-systems. The problem is, however, to define where systems boundaries existed for conceptual purposes, in what overall was an open transnational environment. In effect, there was a series of systems and sub-systems involved from the domestic through to the international levels, with linkages between them. In mapping the overall framework a network of complex communications channels emerges, which enabled information (whatever its initial origin) to disseminate through the systems levels. Such communications are important both in terms of defining the systems network and in terms of their actual content. A transnational model representing the essence of these systems and their links is represented in Figure 8.1.

Two states are represented as decision-systems showing both internal and international linkages with the EMBO – ESF relationship taken to be a system in its own right, the impact of which went beyond Europe.[1] Although it is presented as a two-state model, the simplifications are such that it can be applied as a characterisation of how most of the states involved in

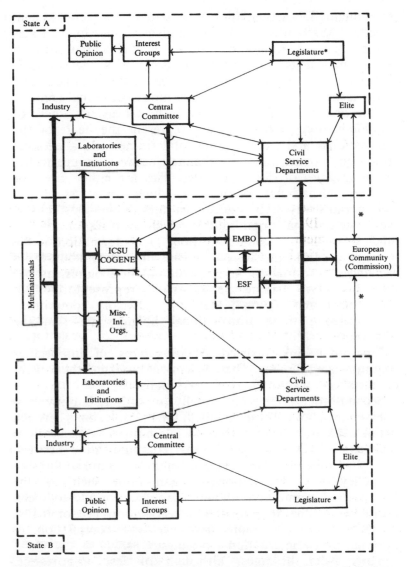

8.1: Simplified two-state communication model for the genetic manipulation issue area

Key: → Main domestic and minor transnational links
➡ Main transnational (and transgovernmental) links
* Where applicable
-- System boundary

recombinant DNA regulation were part of an overall system structure. Much attention has been given to the United States and the United Kingdom because of their importance in setting precedents for others and because of the contrasts between them. But we should not forget all the other countries which faced the necessity of establishing operational controls for recombinant DNA safety. The 'heavier' lines of communication in the diagram represent the heart of the transnational and transgovernmental framework, suggesting established communications channels between groups within the boundaries of different states. Ties are either direct or through international organisations which themselves interact in important ways. The figure illustrates the key positions of these international organisations which were responsible for much processing of communications and information, in the manner described in Chapter 5. Thus individual laboratories exchanged information across national boundaries, as did industry, civil service departments, representatives of government and national regulatory advisory committees. Characterising these transnational linkages we find the following:

First-order links existed where groups of importance communicated directly. These links were rapid and the content of 'messages' was often more personalised.

Second-order links were equally important, but involved intervening actors. These included, for example, links between national advisory committees via international organisations such as EMBO, and via personal communication between individual scientists. Government departments linked through a policy-based international organisation such as the European Community, should also be considered here. Multilateral rather than bilateral considerations predominated in this context. Of note here was the compilation of information by international organisations.

Third-order linkages, although of less importance, nevertheless existed. In these communication occurred with two or more intervening agencies, for example, an NIH policy document obtained by one laboratory, passed on to another, before being passed to a national advisory committee of a different country.

Taking a view that organisations are themselves systems in their own right, we are left with patterns of interaction readily described as multilevel overlapping systems, which preceding chapters have, to an extent, isolated and described.

In a negative sense, however, it is striking that a number of transnational links were weak in relation to the importance of certain domestic groups. Interest or pressure-group activity was not especially noticeable at international levels, despite the fact that these groups could be quite vociferous within individual states. In contrast the scientific community, which is highly international in outlook and has on occasion been characterised as an 'invisible college',[2] was much more suited to developing transnational standpoints and strategies. Common ground on the part of the scientists was more easily identified through the central roles of their international organisations, through attending international conferences, and through personal communications.

Lack of direct contact at the level of governmental elites was also noticeable. This was not too surprising, however, as policy formation and implementation occurred on the whole at departmental levels of the civil service and not at cabinet levels. But in terms of policy-making, both within Europe and across the Atlantic, transgovernmental[3] linkages were of some importance as civil servants kept themselves informed of international developments.

Information, and particularly the active interpretation of information, are critical in situations of uncertainty, perhaps endowing political advantage to certain actors. With good communication links needing a degree of 'tuning' or empathy between sender and receiver,[4] it can be argued that the scientific community, generally speaking, met this requirement. Although it can be questioned whether scientists completely avoid the pressures of nationalism, especially in large-scale prestige science, it is fair to say that for the genetic manipulators the common international fear of excessive regulation enhanced their unity of purpose. This ability and readiness to communicate, in turn, facilitated their sharing of compiled and assessed information. Industry also had a common empathy in as much as industrialists tend to be open to each other's interests as long as they are not directly competitive, for example in pursuing new products or patents. But at the time that guidelines were being developed and introduced the industrial community was more concerned with the maintenance of the necessary freedom to use the new science, to compete subsequently in its exploitation. There were also strong overlaps, in this respect, with their academic counterparts, especially as many of the new start-up companies maintained strong academic links. In sum, decision systems thrive on the availability of information upon which to base choices, and those who used, or hoped to use, the techniques of recombinant DNA were well positioned to capitalise on their direct or indirect access to information channels and decision forums.

If we turn to organisations themselves, it is evident that all possessed formal procedures for collection of information and decision-making. In many cases specialist subcommittees fulfilled different roles, such as: national guideline implementation; transnational risk assessment; guideline comparison; policy and guideline harmonisation; information compilation and dissemination; political consensus formation; arranging international workshops and conferences. Often the activities of particular sub-systems of institutions would make their resources available directly to external bodies; for example, in the case of the Liaison Committee of the ESF and COGENE as a part of the ICSU. Given the overall complexity of the international communication network it is clear that most organisations were 'open' systems with relatively permeable boundaries.

However, the boundary problem is made more difficult by the central roles played by certain individuals in terms of communications paths. As has been mentioned, some individuals, as a survey of the memberships of various committees will show, were active in a number of forums. In many ways this is to be expected when the appropriate administrative expertise and scientific background capable of adapting to the science of the new recombinant DNA techniques were rare, given the relatively few scientists initially active in the field. Thus, membership of domestic central committees could overlap with one or more international committees. Some individuals, therefore, were in strong boundary positions and able to influence the spread of information and the development of policy. John Tooze and Sir John Kendrew were most noted in this respect displaying sensitivity to both the expressions of concern and the potential of the new techniques. In the case of Tooze, this individual clearly had an influence on the development of European attitudes arguing for effective harmonisation through discussion, rather than legislation or a Directive of the form the Commission produced.

Important as it is to be aware of the complex set of overlapping systems at the transnational level and the elaborate communication network involved, it is also very important not to forget the content of the communications in the context of the related transnational politics. Much of the content was in the event *technical* information, centring on, for example, new host-vector systems, proposed experiments, risk assessment exercises, and the sharing of research ideas. The content of different states' guidelines and technical comparisons were commonly available in the international institutions. In addition, the promotion and running of training courses was notable in this type of exchange.

A second area of information content appertained to issues of *policy*. Viewpoints were exchanged over issues such as relaxing or harmonising guidelines, legislative proposals, monitoring and participation. In terms of participation, attention focused on the desirability of the involvement of representatives of the 'public', industry, specialist scientists in disease-related fields, and trade unionists. Policy considerations tended to involve the giving of advice by those compiling reports on international practice, although in the European Community policy discussions included the possibility of imposing constraints. Individual countries could consult other countries' committees or international organisations for advice. Policy choices were therefore made with considerable awareness of the international trends.

For Europe the ESF was a valuable forum to exchange information, as its membership came from 18 countries, and, because Canadian and US observers attended, its influence extended even further. On the whole, though, transatlantic links were more bilateral as far as policy formation was concerned, with British officials, for example, using personal contacts in the HEW, while Shirley Williams, the Secretary of State for Education and Science, maintained informal links with the US Secretary for Health, Education and Welfare, Joseph Califano. Moreover, GMAG members have crossed the Atlantic to familiarise themselves with US procedures.[5]

Of course, it is also evident that with reference to the transnational framework a number of policy questions, argued here to be of importance, were given little, if any, real consideration. For example, little attempt was made to compare formally the issues involved in the recombinant DNA case with general social policy towards risk in technology as a whole, or with other specific technologies (although some brief thought along these lines was apparent in the European Community investigations and within the UK Health and Safety Executive). Part of the reason for this was the way the recombinant DNA case was taken up by the agencies with remits concerned with relatively narrow fields. Biomedical research funding groups often developed the first responses, and became involved in control policy through holding the purse strings. Those agencies with wider remits usually compartmentalised issues and thus established subcommittees specific to genetic manipulation.

A second important omission was a lack of in-depth consideration of the deliberate 'misuse' of genetic manipulation in terms of the construction of biological weapons by states or other political groups. Some mention in passing occurred in the NIH *Environmental Impact Statement* and in the Ashby Report. The rationale given in these reports and elsewhere was that the International Biological Warfare Convention (IBWC) covered the techniques and such issues were really beyond the agencies concerned with basic laboratory research. More subtly, although users of the recombinant DNA techniques could readily comprehend the potential for weapons most wished to avoid the consequences of opening that avenue of debate – namely, further controversy over genetic manipulation. There is a real question about whether or not the IBWC would stop states developing biological weapons if new biological knowledge made their potential utility to the military seem much higher than when the Convention was signed.[6]

Largely avoided by the commanding institutions that showed interest in genetic manipulation, the issues of gene therapy and other human applications of genetic manipulation were not addressed in publicly prominent depth until the 1980s, with, notably, the US Presidential Commission's report entitled *Splicing Life*. The report argued that future developments could not be adequately safeguarded under the current procedures. Ethical problems would be faced in manipulating human genes, especially combinations of human and animal genes. Interestingly, Dr Martin Cline appeared before the Commission and suggested that scientific inquisitiveness might be impossible to regulate.[7]

Nevertheless, although not all issues were addressed in depth, there was a vast quantity of exchange of views internationally, a fact easily demonstrated by browsing through the 1700 or so letters and 11,000 articles collected by 1978 for the Recombinant DNA Collection of the Massachusetts Institute of Technology With certain organisations in important nodal points of the overall system, the 'chain' and 'network' analogies of the Evans model of interorganisational systems are appropriate. So too is his reference to the importance of boundary personnel. However, the Evans model and many functional assessments of organisations can overemphasise structural and communications-based interpretations of events. In the case of genetic manipulation there were overtly political elements as well. That is, 'political' in the sense of values in conflict. This was emphasised by the scientists themselves learning the basics of practical politics when, for example, they took on the would-be legislators. Beliefs and wishes at times even became partial substitutes for empirical evidence. Local issues blazed with the flames of politics in places like Cambridge, Massachusetts. Competing interests were evident when, at the Wye Conference, US scientists tried, in Sir John Kendrew's words, to 'steamroller' those in the UK into accepting lower guidelines. Political divisions were also important when there was difficulty in agreeing on issues like the European Community Draft Directive and funding of research in the field. National representatives raised various objections to individual proposals. On the face of it these represent some of the more obvious examples. Is there a case, therefore, to suggest that at the heart of the recombinant DNA debate were, what Steinbruner termed, 'complex decision problems'?

It has been suggested that complex decision problems involve values which are affected by choices taken in a zero-sum trade-off sense. Thus one set of values is satisfied at a corresponding cost to other sets. There is, of course, the added question of whether or not any such conflict is founded upon very real differences of view, or simply misperceptions about the conflict-situation held by the participants. In effect, otherwise positive-sum outcomes might be forfeited in ignorance, or through uncertainty. Values were in perceived conflict over genetic manipulation in the heady days of the 1970s – that is undeniable – but we need to look at the decision processes more closely.

Scientists themselves at Asilomar II and elsewhere expressed the need to examine in detail 'all of the issues'. There may have been a loose consensus at Asilomar, but it became apparent that not all of the international assemblage of scientists were agreed on the degrees of risk which could be conjectured and the actions which should be taken. Interest groups took up the discussion drawing on these divisions, called for openness, for further pause, and for legislation. Government departments which were involved held different views from those of scientists concerning the imperatives for work to progress as rapidly as possible, and administrators perceived the requirements of submitting protocols or memoranda of understanding as less disruptive than scientists and industrialists. A very important value held by some groups was purely and simply the right to participate in the making of decisions, without prejudice to whatever decision that might emerge. Issues over which decisions brought forward conflicting values therefore included:

Should the work continue (that it should was an assumption for many)? What degrees of containment were necessary for what experiments? What type of monitoring mechanism was necessary? Where should the authority lie? Who should participate in decision-making? How could industry be equitably covered? Would legislation help – and if so, in what form? How much risk assessment work would be sufficient to underwrite guideline relaxation? How much relaxation would be acceptable at any particular time? How much national effort should be put into promoting the new components of biotechnology? How best could international harmonisation be achieved? What parameters of inquiry should exist in examining the likes of genetic manipulation? What about broader ethical, moral and religious issues? These questions, and others, were often inseparable from underlying value stances. For example, different views of what 'freedom of science' meant were involved. There was the question of whether or not the 'social responsibility' of science, its responsiveness to society, and its accountability should be emphasised. Views were held on the legitimacy of curtailing activities which might cause harm to others, in turn leading into debates about social gains measured against social risks. Others held to the value of defending open competition between companies, states and other institutions in relation to overall benefits which could accrue. Even the 'freedom of the press' became an issue, in relation to the Wye Conference in particular, while others defended the rights of broad public participation, including their rights to make erroneous decisions for the sake of caution.

Much of the basis for values coming into conflict was, therefore, concomitant with the general level of uncertainty involved, while the credibility of scientists, as a group, was challenged further by some misuses of information and by guideline violations. In terms of undertaking risk assessment or risk-benefit assessment, the specific problems of uncertainty have already been outlined. At best, attempts to be rational occurred within certain limitations. 'Bounded' or 'limited' rationality better describes the incremental pattern of decision-making evident throughout the transnational milieu. This resulted in satisfactory rather than optimal decisions.

An example of limited rationality can be seen in decision-making over the form and content of the US guidelines. Lack of comprehensive information led to a search for suitable criteria for allocating safeguards. The limits of rationality were particularly evident with the final recommendations of the RAC for the 1976 US guidelines, where three sets of draft guidelines were examined in parallel. Paragraphs were considered for no more than ten minutes and votes were taken on proposals. Controversial differences had led to the various drafts which the RAC compared and recommended upon. A degree of rationality may have applied in attempting to make an optimum choice, but the nature of the decision process limited it. In organisational terms bounded rationality is the norm where complex problems are compartmentalised on the basis of specialisms and functions. Thus the final coordinator of advice from organisational subsections may have been well outside their individual deliberations. More generally, with the intense lobbying and discussion surrounding most legislative efforts, from the local, through national to international levels, resolutions could at best be satisfactory. Diverse 'power'[8] held by the opposed actors involved precluded ostensibly rational solutions. Similar problems were evident in the establishing of guidelines in most countries that developed their own. In essence, conflicting values are not readily amenable to rational 'trade-off'. Yet, despite this, but in differing degrees in different states, deference was made to the dominant role of the 'expert'.

Uncertainty was a central problem for all concerned. But at least organisations, in many contexts, have displayed procedures to attempt to deal with it. The literature on decision-making procedures suggests that uncertainty leads to the response of organisations trying to avoid or bypass it. This is not as much a criticism as an observation on decision strategies. Short-run decisions were very evident in the case of recombinant DNA. Guidelines were designed as cautious, such that, as Brenner suggested, future revisions would be downwards. Revision was to be frequent to accommodate both new information and short-run feedback from scientists using the containment facilities and acquiring empirical evidence. However, not all feedback resulting from early decisions and recommendations from Asilomar, the RAC, the Williams working group, and elsewhere was positive and reinforcing from the point of view of the scientists who drew attention to the conjectured hazards. Although well thought out, the temporary measures outlined in dramatic fashion by the Berg letter and the Asilomar meeting led to an upsurge of press and public interest. This reaction in turn led to an abortive attempt to draft a second 'Berg' letter, almost four years after the first, this time summarising reasons for relaxing early fears.[9] By then, the threat of legislation was at its height. Paradoxically, many considering legislation on their part tried to build clauses into their proposed bills limiting the lifetime of the legislation to perhaps two years, in recognition of the need for future reappraisal. The European Community Draft Directive similarly proposed a regularised revision of the situation.

Aware of the problem of past uncertainty, some of those advocating guideline relaxation tried to show that the scientific community was now nearly *certain* in its view that genetic manipulation was safe, following the risk assessment activity discussed and the failure of the application of the techniques to demonstrate hazard. In the wider context, however, many interested non-scientists could not readily adapt to the transition from expressions of uncertainty, leading to incremental decision-making, to claims of near certainty in stressing the safety of the work (given precautions).

All in all, the scientists initially acted in a way that would allow subsequent information to alter tentative decisions taken. Their mistake was in failing to appreciate the much wider politicised responses coming from uncertain non-scientists. On the other hand, many institutions, including administrative organs, responded by recourse to implementing standard procedures. Committees of investigation made recommendations which were sent to a higher level for approval and which were then implemented by adapting existing mechanisms. The overall process, observed transnationally, was for uncertainty to be played down where possible or contained within incremental decision procedures. This process was transnational in that decisions in different locations were taken on the basis of shared information and with awareness of precedents set elsewhere.

It is all very well arguing that decisions were based on incremental procedures, which is quite understandable in the circumstances described, but we must not forget the associated question of the search for alternative courses of action. Decisions were taken within systems of interaction displaying different degrees of 'organisation'. Berg's small committee informally consulted colleagues and made one-off decisions through their letter and the Asilomar II conference. They had few rules to go by. At the other extreme, consensus pressure for guideline relaxation was applied by a *transnational community* of scientists, bound together by common interest. Such output from a loose system does not really reflect organisation. Yet many institutions or organisations were active in decision-making, but not always addressing many alternatives. The charges made against the Office of the Director, NIH, and the limited consideration of alternatives in its *Environmental Impact Statement* following the first US guidelines, could well have been applied to other institutions. It is clear that nowhere was it seriously contested that genetic manipulation should continue, or at least suggested that it be greatly curtailed in the long run. Within states and internationally, decision alternatives were developed within parameters of what was acceptable to the community involved. This was most noticeable in terms of voluntary guidelines. If they were too strong, the 'threat' was of limited compliance with them. The *raison d'être* of some influential organisations such as the ICSU, EMBO and the ESF was the promotion in general of scientific study and the exchange of information. Wider issues were not their normal concern and although they were aware of the potency of genetic manipulation to stir up ethical and philosophical questioning, they were not keen to embrace the debate directly. When wider policy searching did occur it was largely confined within some institutional structures such as those examining legislation, within organised interest groups, and within certain local forums such as universities or local government offices.

In sum, large-scale risk-benefit assessment was never applied to precedent setting policy decisions. Within the communications networks, the fact that information was provided or collated by like-minded groups reinforced the overall limited selection of policy options. Indeed, there is evidence that some groups actively tried to neutralise considerations of certain options by strongly supporting an alternative, for example researchers supporting relatively lax legislation proposals to avoid the more extreme. Rather similarly, many scientists had earlier supported the development of guidelines to offset any legislation. In part it was the international build-up of competitive momentum to get on with the work, the acceptance of similar restraints in the many states, and the general likemindedness of the many participants that kept the parameters of policy search relatively narrow around the world.

In all decision-making situations those who participate possess some degree of influence or power. In effect participation is an important element in any definition of power. Although analysis of the domestic processes of decision-making in issues surrounding genetic manipulation may reveal some characteristics of elite structures, at a transnational level the politics of pluralist participation were more important. A corollary of the notion of power lying in participation is the act of excluding other groups. Thus images of organisation inferring the 'mobilisation of bias' and the possibility of 'non-decision-making' deserve attention when looking at the range of actors involved.

Professional decision-makers in departments responsible for domestic policy were, from the beginning, in positions to influence procedures. However, the conjectured hazards were announced by scientists in an authoritative but technical way and the administrators' position was really one of response to a request to consider the issues further. The result was working parties in many states calling witnesses and drawing up reports and recommendations. Guidelines were then drafted by scientists and administered by departmental professionals, scientists and local institutions. Requests to participate in this decision process came from numerous quarters. The lines of contention settled on the difference between those who wished the process to remain one of technical assessment, utilising the knowledge of the expert, and those who wished for a process of decision-making commanding greater legitimacy. It is clear that environmentalists, industrialists, trade unionists and others wished to contribute to major policy decisions. US environmentalists wrote many letters to the NIH and the RAC suggesting policy and emphasising caution. Industry lobbied bureaucrats over the European Community proposed Directive. In Japan interdisciplinary assessment of genetic manipulation was encouraged. In the UK trade unionists successfully campaigned for representation on the planned GMAG, while in the Netherlands lobbying by left-wing political parties brought a government response such that Dutch guidelines were probably the toughest in the world.

Although the issue of participation was much publicised in the United States, it was generally less of an issue in Europe and the United Kingdom. In part, the smaller scale of recombinant DNA research activity in many states made the question less obvious, and partly the traditional openness of the US political system facilitated lobbying at local and national levels. Consequently, in the US, the environmentalists achieved some influence with regard to legislation and local regulations, although assisted by some sensationalist press coverage. On the other hand, the UK model was established to include wider participation from the beginning. In recommending the creation of GMAG the Williams Report argued:

Since a central advisory service will need to command the respect of the public as well as of the scientific community, including scientists in industry, the membership of the GMAG should include not only scientists with knowledge both of the techniques in question and of the relevant safety precautions and containment measures but also able to take account of the interests of employees and the general public.[10]

Ironically GMAG was to be very secretive in its operations, and additionally some problems arose over which trade unions should be included and who should represent the public interest. France went further after its early public and press controversy, in establishing two national committees, one similar to GMAG and one charged with examining ethical, moral and wider social issues. Japan likewise induced broad participation. Altogether at least eleven states by 1980 had public interest representatives on their national committees. Even the United States moved the composition of the RAC more towards wider participation. Therefore, whilst organisations clearly reflected natural biases which often bordered on political postures regarding the key issues, some institutional effort was also apparent in a number of states to widen participatory input, if sometimes only after campaigning by those previously excluded.

Without overstressing the point, while participation tended to become broader over time and with the increased public profile of genetic manipulation as a technology, many important precedents were set within very narrow parameters. Appeals to the freedom of science and the authority of the expert ensured prudent safety measures, but also the continuation of the work with little high profile attention drawn to issues other than safety. Administrators showed little interest in going beyond the more limited issues of health and public safety while governments remained somewhat distanced. Because of the sophistication of the science involved, openness could be practised without undue adverse publicity – provided the decision-making procedures were deemed legitimate. To assist this the press could be labelled 'ill-informed' or 'sensationalist' and scientists could urge their colleagues to be careful of what they said. Both of these factors were notable in the run up to the Wye Conference. Whether such general observations can be taken to provide support for a view that many 'non-decisions' were involved is debatable. However, regarding genetic manipulation's potential for biological warfare, then biases did clearly restrict the scope of study in those scientific institutions which possessed the knowledge to assess the possibilities, but chose to downplay the issue.

The whole question of participation can finally be summarised with reference to the concept of legitimacy. Much of the conflict endemic in the many debates over genetic manipulation was arguably misconceived due to problems of perception and to failures of legitimacy. Scientists, for example, had brought their conjectured fears to international attention after serious deliberation. They expected, in line with their perceptions, to receive a sober and analytical response, investigating the risks and developing appropriate measures of caution. Scientists fully expected to remain in the central position, discussing internationally and assessing risks perhaps under the auspices of governmental departments. Calls for 'rationality', it has been said, were common. At Asilomar II the realisation dawned that they would have to take actions rapidly or find themselves the victims of their own authoritative statements. Legislation, turmoil and harsh constraints might follow. Conflict between participants in the overall debate to a large extent focused on two levels of uncertainty:

1. It was quite apparent that scientists were divided on the nature and the degree of hazard, despite appearances at Asilomar.
2. Non-scientists appreciated the authoritative nature of the scientists publicising their conjectural fears, but did not understand the science. Their uncertainty was further enhanced by the divisions among scientists and their subsequent actions.

The resulting conflict was compounded by the time-lags involved before public, interest group and legislative actions emerged. By the time non-scientists had found their feet in the technicalities of the new science (or found scientists to support them) and had begun to understand the nature of the risks postulated, some scientists were making new discoveries. The transition of the scientists' views towards less risk was thus underway. But at this point some scientists also began to question their earlier actions, something which to non-scientist interest groups seemed a turn around in the face of regulations and scrutiny. From such beginnings the misperceptions and the outright politicking took root. Scientists wished to retain authority, but could not understand the political requirements of legitimacy. Bureaucratic imperatives and vocal criticism from interest groups widened the calls for participation and much of the value differences became apparent at transnational levels as the whole context of the debate became more complex. The demands of legitimacy and the politics of participation thus came down to the old maxim that not only should things be done, but they should also be seen to be done.

SUMMARY

In the Introduction attention was drawn to the sorts of underlying questions which have motivated this study. By way of summary it is worth returning to them.

To what extent were there political biases in the development of safety guidelines and their implementation? An argument presented throughout has been that the genetic manipulation debate was politicised because of the clashes of values, interests and perceptions involved. It is clear that governments, scientists, and professional and international organisations all tried to limit the political context of the debates over the technology. Full open forums, such as public inquiries, were not usually seen as relevant by those who dominated the decision-making processes, although in the United States many open meetings and legislative hearings were held. The emphasis on the role of the expert maintained a strong influence on the part of the scientists as a group (even if a minority of scientists were particularly cautious).

At the governmental and transgovernmental levels political will was not sufficient to achieve a policy of international harmonisation of regulations. Pressures from domestic and transnational groups reinforced the move away from imposing harmonisation in the wider context of their perception of declining risks and their wish to deregulate. A number of states never developed provisions for ensuring all users of recombinant DNA techniques were equally covered, in part because of the resistance to legislation. Thus successful lobbying combined with new information constrained the provision of universal safeguards. Important in this was the ever-increasing industrial exploitation of genetic manipulation and biotechnology. By the time that regulation of genetic manipulation was being reduced substantial industrial benefits were on the horizon, and governments were active in nurturing biotechnology in their countries. Even if greater risk had manifested itself it is likely that the perceived importance of the industrial promise of biotechnology would have ensured rapid exploitation of the new knowledge. After all, the early regulations had not significantly held up the work despite the protests from some quarters.

Although the uncertainty underlying decision processes led to incremental decisions there were also the inevitable compromises where values and interests clashed. Both the legitimacy of the decision processes and the specific results of decision taking in various states came under question by different groups with their different objectives. Most states, for example, faced at one time or another criticisms that their guidelines were too harsh relative to those of other countries. If controversy caused stagnation in the effective production of comprehensive guideline cover then it is of luck that the risks were not realised as most states had limited facilities for enforcement of guidelines and inspection.

Was the range of options considered in the important centres of decision-making unnecessarily narrow? In retrospect many control options could be postulated. Different emphases might have been placed on inspection, central regulation, range of coverage, punishments, and so on. Or even a complete ban might have occurred. The biases discussed were undoubtedly of importance here. But there were other limits on the search for options arising because of the limitations on information available due to uncertainty. The initial responses to the conjectured hazards were transnationally incremental. Thus, voluntary suspension of work gave way to tentative but quite restrictive Asilomar guidelines. These in turn prompted more formal controls to which modifications were made as new information became available. Under uncertainty such incrementation is creditable given the difficulties of comprehensive rationality. The real problem was one of legitimacy over who should be involved in the decision process. Political factors were more apparent in the way that guidelines were relaxed than in the way that they were initially introduced. This point may be taken further. For all the limitations that may have existed, the case of genetic manipulation was nevertheless unprecedented in the extent that serious consideration was given to potential risks long before any were realised. The nuclear industry suffered decades of controversy following the initial secrecy of its development and the blatantly inadequate understanding of the consequences of human exposure to radiation. In the UK the nuclear establishment at Sellafield (formerly Windscale) is only belatedly embracing the concept of openness, already to evident good effect in terms of its public image. Yet, biotechnology in its turn must not now begin to act rashly in denying all risk and therefore undoing the generally sensible approach to risk it had shown, despite any problems involved in that. The evidence of the way that pressure came to bear in bringing guidelines down may, however, raise a question mark in this respect. At this stage it is still appropriate to continue the public scrutiny of biotechnology because of remaining hazards and because of the impending environmental exposure of biotechnology products. Commercial gains must not cloud future judgement about hazard.

Were safety controls searched for within existing operational frameworks on the basis of an assumption that the frameworks themselves should not have to accept more than minimal change? Because of uncertainty and incremental decision-making there was a tendency to avoid major changes to existing frameworks. As far as the early stages of response to the conjectured hazards are concerned this is not a criticism. Those who recognised the potential risks not unnaturally expected to continue to examine the issue and they and those they alerted preferred to operate within institutions with which they were familiar. With such a shortage of expertise in the new science these people were important to the subsequent involvement of administrators and working parties.

Problems of utilising existing systems really only emerged when it was decided to end the moratorium and allow the work to continue under guideline constraints. Administration of guidelines by agencies sponsoring research in many countries led to charges of conflicts of interest. In this context there is a strong case for saying that regulation should be separated from organisations whose normal bias would be towards protecting and encouraging science activity *per se*. Such early action could have avoided the problems of not all researchers being accountable to guidelines and perhaps enhanced legitimacy. The nuclear industry has faced similar charges of conflicts of interest where agencies concerned with its control and safety have also been active promoters. Yet, with genetic manipulation, the alternative of finding resources and people to establish new structures to cover what was after all conjectured risk would have also been difficult to justify. In many ways the outcome was a compromise. Institutional innovations were introduced but to a limited degree. Thus the practice of having a national committee to provide advice or to draft guidelines was common, supported by the

encouragement of risk assessment. Overall, the international trend was to use existing structures as far as possible, but with adaptation in an incremental fashion.

Was there any concerted effort to try to apply the lessons of other technologies regarding safety issues and legitimacy in the development of policy? If there is to be one general overriding criticism of the procedures applied to investigating the risks of genetic manipulation and the development of safeguards, it is that little effort was really made to compare the lessons of other technologies regarding the management of potential hazards. There might have been much to learn by calling in more specialists in hazard control *per se* and experts in specific if unrelated hazards. Procedures to engender confidence, to operationalise risk assessment and gain new insights could have emerged. This criticism is not merely directed against recombinant DNA, but against the general tendency of societies to compartmentalise such issues rather than to centralise all hazard and potential hazard policy. If this had been the case a more appropriate agency might have existed for assimilation of the problems. Susan Wright has criticised the failure of the United States to use what provisions that did exist.[11] In the UK it became apparent that although the Health and Safety at Work Act could be used, it nevertheless required a redefinition of 'work'. The UK was fortunate, however, in having the HSE, although this comprised only part of the institutionalised process. Yet some comparison was made with dangerous pathogens in general within the UK, the WHO, and elsewhere. Japan was the more exceptional in locating recombinant DNA issues within more interdisciplinary studies of technological impact on society, although their efforts were not of great impact beyond their borders. In looking to the future, there is a case to be made that new technologies of quite diverse content may give rise to overlapping problems related to the needs of social choice, safety and accountability. Added to this is the necessity of maintaining vigilance over existing technologies. The availability of information to interested parties may even represent a fundamental step in this direction. In this instance the success of the US Congress Office of Technology Assessment, which services the politicians by providing impartial assessments of new technological developments, can be contrasted with the growth of requests from some quarters for something similar in the UK.

How effective were the communications between the many national, transnational and international groups? The argument presented in this chapter is that on the whole the communications network between the many actors involved was extensive, and efforts have been made to conceptually map the main links. It is worth, however, repeating that the links between the critics of DNA were weak at the international level compared with the scientists and the industrialists. Nevertheless, the content of debates within the most important states (in terms of their significance in the field of genetic manipulation and microbiology and consequently their guideline attitudes) was well reported internationally, such that concerned groups were well informed. Transgovernmental exchanges were also notable as many countries looked to the policies of the US and the UK and others for inspiration.

The international ethos of business and science has helped them in keeping abreast of new technological developments and their potential applications. In association with what is effectively an assumption that 'progress' in technical fields is desirable, this ethos helps determine the value-stance taken by such groups whenever a new technology raises questions which invoke value judgements. This was no exception in biotechnology. Recognition of potential risk is one step, but the second step is the determination of what level of risk in an activity is socially acceptable. Those who explicitly stand to gain may display a relatively higher threshold of what they view as acceptable. Watson may have recommended the attitude of get on with the work and risk the law suits if anything went wrong, but this does not assist in the sensible exchange of views over fears where the non-expert may lack the scientist's sense of working at the forefront of human knowledge. It is most unfortunate that misperceptions between different groups, combined with their alternative value-sets, undermine constructive communication between them. We all perhaps lose out as a result.

Were there any particular problems at the international level?

Although the European Community embarked upon the proposal that member states should harmonise their guidelines and implementation procedures, it should be said that with the diversity of national requirements and experiences genetic manipulation was not likely to have brought enough political compromise for universal harmonisation. Had the issue been amenable to rational solution then harmonisation might have resulted. But given the range of views involved and the different national attitudes to promoting new technologies this failure is understandable. This is not to say that the overall degree of safety provided for was not similar, but that the approaches to the problem were different. Yet a serious consequence of the national differences was the 'lowest common denominator' effect when the guideline revision processes were underway. Competition to exploit the promise of recombinant DNA plus the lowest common denominator effect may have brought untoward pressure for relaxation before the evidence for lowering restrictions was conclusive.

Biotechnology has not been the first or only technology to have experienced problems akin to the 'lowest common denominator' effect. It also may not have seen the last of the problem either. Commercial pressure in the form of the profit motive and the competitive search for new markets has often led industry into positions where countries are chosen for new direct investment where their regulations are, internationally, comparatively lax. It is a sad, if true, fact that many developing countries cannot afford the luxury of comprehensive environmental policies in respect of industrial and technological activity. With the lower production costs which ensue, the multinationals may locate in certain countries and accept that they will cause more pollution or pose higher risks there than elsewhere. This point was brought home when the chemical disaster at Bhopal raised the question of whether demands for compensation would have been higher had the disaster occurred in Union Carbide's similar plant in the US. The nuclear industry as a low-probability, high-consequence risk industry has long faced such difficulties in terms of differing national standards. Pringle and Spigelman put this point succinctly in reference to a discovery in 1977 by the International Atomic Energy Agency that:

sales by Western reactor companies to less-developed nations were frequently of a less-safe design and had less-rigorous operating specifications than were required in the exporting countries. The IAEA report ... focused particularly on special models that had been made to fit specific requirements and had never been subject to regulatory review.[12]

Biotechnology in the future must not generate further practices of this sort as more and more states become keen to develop expertise in the field. Developing states will hopefully not become the testing-ground for environmental releases of new organisms or plants because they might be disallowed in some developed states. Similarly, in the development of new medical practices it is to be hoped that the Cline lesson is learnt and researchers do not embark upon deliberate searching for countries which will allow work on human DNA, for example, which is banned elsewhere. The ethical and moral dilemmas may need much more thought before certain steps are taken.

In a broader sense, biotechnology, after its revolutionary thrust forward in recent times, will continue to raise issues of international importance. DNA represents the 'stuff' of life, and cannot be interfered with indiscriminately. There is such a thing as 'dangerous knowledge', or at a minimum controversial knowledge. It is with caution that we must proceed in the exploitation of our increasing understanding of the language of life. On balance we have adapted well to the hazards posed by the basic manipulation of DNA in microorganism, host – vector systems. However, we must keep the warnings of Huxley in mind as we head towards the twenty-first century.

Notes

INTRODUCTION

1. Aldous Huxley, 'Foreword' (1946), *Brave New World*, Penguin, Harmondsworth, 1932, pp.9–10. Huxley noted that in his book the only scientific advances he specifically described were those involving the application to human beings of the result of future research in biology, physiology and psychology.
2. See, for example, J. Glover, *What Sort of People Should There Be?* Penguin, Harmondsworth, 1984.
3. Discussed in Chapter 2.
4. A.T. Bull, G. Holt and M.D. Lilly, *Biotechnology: International Trends and Perspectives*, Organisation for Economic Cooperation and Development, Paris, 1982, p.21. Eleven previous definitions are produced in an appendix to this report and are taken account of in their definition.
5. See, for example, M. Rogers, *Biohazard*, Knopf, New York, 1977; J. Goodfield, *Playing God*, Hutchinson & Co., London, 1977; N. Wade, *The Ultimate Experiment: Man-Made Evolution*, Walker & Co., New York, 1977; J. Lear, *Recombinant DNA: The Untold Story*, Crown Publishers, New York, 1979; R. Hutton, *Bio-Revolution: DNA and the Ethics of Man-Made Life*, New American Library, New York, 1978; C. Grobstein, *A Double Image of the Double Helix: The Recombinant DNA Debate*, W.H. Freeman & Co., San Francisco, 1979; D.A. Jackson and S.P. Stich (eds), *The Recombinant DNA Debate*, Prentice-Hall, Englewood Cliffs, NJ, 1979; E. Yoxen, *The Gene Business: Who Should Control Biotechnology?* Pan Books, London and Sydney, 1983; J.D. Watson and J. Tooze, *The DNA Story: A Documentary History of Gene Cloning*, W.H. Freeman & Co., San Francisco, 1981; S. Krimsky, *Genetic Alchemy: A Social History of the Recombinant DNA Controversy*, MIT Press, Cambridge, Mass., 1982.
6. See M. Sittig and R. Noyes, *Genetic Engineering and Biotechnology Firms Worldwide Directory 1983/1984*, Sittig and Noyes, Kingston, NJ, 1984.

CHAPTER 1

1. J.D. Watson and J. Tooze, *The DNA Story: A Documentary History of Gene Cloning*, W.H. Freeman & Co., San Francisco, 1981, p.vii.
2. With its first chair established in 1919 at Aberystwyth.
3. International Political Economy would suggest important linkages in a multidimensional fashion between *all* of the following: domestic economics, domestic politics, international economics and international politics.
4. The broad category of techniques used to produce what are known as recombinant DNA molecules perhaps engender different perceptions when labelled 'genetic engineering' or 'gene splicing' than when labelled 'recombinant DNA techniques', 'genetic manipulation' or when in conjunction with other techniques are termed 'biotechnology'.
5. Late nineteenth-century biology had already identified the role of chromosomes and had determined that traits were controlled by hereditary factors linked to chromosomes in some collective fashion. The concept of what we call a gene was therefore known, but identification of the biochemical importance of DNA is the real starting point of significance here. The above provides a very brief summary of this. See S.R. Kushner, 'The Development and Utilization of Recombinant DNA Technology', in J. Richards (ed.), *Recombinant DNA; Science, Ethics and Politics*, Academic Press, London, 1978, pp.35–58. See also Watson and Tooze, *op. cit.*, pp.529–83. For a general description of the discovery of genetic functions through to the progress of biotechnology see S. Yanchinski, *Setting Genes to Work: The Industrial Era of Biotechnology*, Penguin, Harmondsworth, 1985.
6. J.D. Watson and F.H.C. Crick, 'Molecular Structure of Nucleic Acids. A Structure for Deoxyribose Nucleic Acid', *Nature*, Vol. 171, 25 April 1953, pp.737–8. See also J.D. Watson and F.H.C. Crick, 'Genetic Implications of the Structure of Deoxyribose Nucleic Acid', *Nature*, Vol. 171, 30 May 1953, pp.964–7. For an interesting narrative of the story of how Watson and Crick made their discoveries see J.D. Watson, *The Double Helix*, Penguin, Harmondsworth, 1968.
7. These free bases include three of the DNA bases, adenine, cytosine and guanine, but with thymine replaced by a further base called uracil.
8. In the surrounding cytoplasm where amino acids are readily available, segments of a second form of RNA, known as ribosomal RNA, become chemically attached to individual amino acids. Having attached itself to a particular amino acid this RNA is able to 'read' the genetic alphabet of the messenger RNA and deposit its amino acid at the appropriate point relative to the other amino acids being located by more ribosomal RNA. A chain of amino acids results. For quite superb illustrations of this whole process see Watson and Tooze, *op. cit.*
9. C. Grobstein, *A Double Image of the Double Helix: The Recombinant DNA Debate*, W.H. Freeman & Co., San Francisco, 1979, p.11.
10. When the DNA code is 'read' it is read in different directions along each backbone as the chemical structure of the two backbones run in

opposite directions (often called polarities). Thus in the example here the cut in the two backbones is in effect at an identical place given the opposite polarities.

11. For tables of various restriction enzymes and the sequences they recognise, see Watson and Tooze, *op. cit.*, p.554.

12. The method described here is perhaps the most common approach to manipulating DNA in the laboratory, but it is not the only one. An alternative involves the use of an enzyme called *terminal transferase* which has the ability to add either As or Ts one at a time. By adding As to the end of one segment of DNA to be joined and Ts to the end of the other segment, it would be possible for them to bond given the propensity of A and T to pair. Many experiments might indeed involve combinations of these methods.

13. See D.A. Jackson, 'Principles and Applications of Recombinant DNA Methodology', in D.A. Jackson and S.P. Stich (eds), *The Recombinant DNA Debate*, Prentice-Hall, Englewood Cliffs, NJ, 1979.

14. Indeed the work of Cohen and Boyer was to become the subject of a decade of legal wrangling over the patenting of their recombinant DNA techniques. The patents eventually awarded were to be held by their respective universities, Stanford and UCLA.

15. Eukaryotes cover a range from relatively simple organisms to primates, including human cells.

16. See 'Machines to Read Genes', *Economist*, 24 August 1985, p.81. See also 'Banking DNA Sequences', editorial, *Nature*, Vol. 285, 8 May 1980, p.59.

17. Such as cell-hybridisation, or the amalgamation of the nucleii of two different cells, thus mixing genetic characteristics.

18. See A. Campbell, 'Natural Modes of Genetic Exchange and Change', in J. Morgan and W.J. Whelan (eds), *Recombinant DNA and Genetic Experimentation*, Pergamon Press, Oxford, New York, 1979, pp.21–7.

19. See J.D. Steinbruner, *The Cybernetic Theory of Decision: New Dimensions of Political Analysis*, Princeton UP, Princeton, NJ, 1974, p.16.

20. For an interesting and provocative discussion of the possibilities for redefining zero-sum situations of conflict in positive-sum terms, see J.W. Burton, *Deviance, Terrorism, and War: The Process of Solving Unsolved Social and Political Problems*, Martin Robertson, Oxford, 1979.

21. For a general discussion of rationality as addressed in a number of fields see A. Heath, *Rational Choice and Social Exchange*, Cambridge University Press, Cambridge, 1976. See also A.G. McGrew and M.J. Wilson (eds), *Decision Making: Approaches and Analysis,* Manchester UP, Manchester, 1982, and, especially, M. Carley, 'Analytic Rationality', pp.60–6.

22. See M. Carley, ibid., p.61.

23. For a discussion of the nature of values, their relationship with norms and policy-making, see G. Vickers, 'Values, Norms and Policies', in, F.G. Castles *et al.* (eds), *Decisions, Organizations and Society* (2nd

edition), Penguin, Harmondsworth, 1976, pp.129–42. On the psychological elements of decision-making see Steinbruner, *op. cit.*

24. For a discussion of the level of analysis problem in international relations, see J.D. Singer, 'The Level-of-Analysis Problem in International Relations', in K. Knorr and S. Verba (eds), *The International System: Theoretical Essays*, Princeton University Press, Princeton, NJ, 1961, pp.77–92.

25. See R.O. Keohane and J.S. Nye (eds), *Transnational Relations and World Politics*, Harvard University Press, Cambridge, Mass., 1972. See also M. Smith, R. Little and M. Shackleton (eds), *Perspectives on World Politics*, Croom Helm, London, 1981, especially section 2.

26. J.W. Burton, *World Society*, Cambridge University Press, Cambridge, 1972, p.46.

27. Much as in probability investigation. An apple is a member of the set of apples, as a pear is a member of the set of pears, but both are members of the larger set of fruit. See J.W. Burton, ibid., p.48.

28. See K.W. Deutsch, *The Nerves of Government: Models of Political Communication and Control*, Free Press, Collier-Macmillan, New York, London, 1963.

29. Deutsch formally defines feedback as meaning 'a communication network that produces action in response to an input of information, and *includes the results of its own action in the new information by which it modifies its subsequent behaviour.*' K.W. Deutsch, ibid., p.88.

30. See R. Cyert and J. March, *A Behavioural Theory of the Firm*, Prentice-Hall, Englewood Cliffs, NJ, 1963.

31. F.E. Kast and J.E. Rosenzweig, *Organisation and Management*, McGraw-Hill, London, 1974, p.415.

32. See Chapter 3.

33. For an analysis of scientists as transnational actors attempting to increase their influence over governments and international government organisations, regarding science policy, see D. Crane, 'Transnational Networks in Basic Science', in Keohane and Nye (eds), *op. cit.*

34. G.T. Allison, *Essence of Decision: Explaining the Cuban Missile Crisis*, Little, Brown & Co., Boston, 1971, p.77. See also Steinbruner, *op cit.*, p.66.

35. See H.A. Simon, *Models of Man, Social and Rational*, John Wiley, New York, 1957, and H.A. Simon, 'Theories of Decision-Making in Economics and Behavioural Science', in Castles *et al.* (eds), *op cit.*, pp.30–48 (excerpts from an article originally published in 1959).

36. See W.M. Evans, *Organization Theory: Structures, Systems and Environments*, John Wiley, London, New York, 1976.

37. Some important approaches relevant to political and sociological analysis have been encapsulated as the debate between elitist and pluralist conceptions. A very useful summary of the debate has been compiled from original papers by Castles *et al.* (eds), *op cit.* In particular, see the contribution by P. Bachrach and M.S. Baratz, 'Two Faces of Power', pp.392–404. The elitist view holds that every human institution involves relatively stable power structures related to its

internal stratification. Dominant groups can thus be recognised where interests regularly prevail. Pluralists see power as participation in decision-making reflecting actual activity rather than elite reputations of power. Exchange theorists suggest power derives from possession of 'resources' others require and cannot readily obtain elsewhere, or from social reciprocity. On the latter see P.M. Blau, *Power and Exchange in Social Life*, John Wiley, New York, 1964, and Heath *op. cit.* Note that Keohane and Nye considered that an effect of transnational relations would be the fostering of transnational pluralism. See note 25.

38. E.E. Schattschneider, *The Semisovereign People*, Holt, Rinehart & Winston, New York, 1960, p.71. Emphasis by Schattschneider.

39. See Bachrach and Baratz, *op. cit.*, and P. Bachrach and M.S. Baratz, *Power and Poverty: Theory and Practice*, Oxford University Press, London, 1970. See also J.W. Jenkins, 'The Case of Non-Decisions', extract from *Policy Analysis*, Martin Robertson, Oxford, 1978, reprinted in McGrew and Wilson (eds), *op. cit.*, pp.318–26.

40 See J.W. Burton, *Deviance, Terrorism and War, op. cit.*, pp.140–56.

41. For a description of this work, see J. Dorman, 'History as She is Made', *New Scientist*, 10 January 1980, pp.86–8.

42. E.H. Carr, *What is History?*, Penguin, Harmondsworth, 1961, p.11.

CHAPTER 2

1. For an interesting and concise summary of this view of the way by which theory is developed see K.N. Waltz, *Theory of International Politics*, Addison-Wesley, Reading, Mass., London, 1979, pp.1–17. Waltz takes the line that theory is not about the discovery of 'laws', but rather is about the explanation of discovered regularities. Theory in this sense he suggests relies on both induction and deduction brought together with creativity.

2. See C. Grobstein, *A Double Image of the Double Helix: The Recombinant DNA Debate*, W.H. Freeman & Co., San Francisco, 1979, p.16. Generally, a number of journalistic style accounts have proved useful as background material. See M. Rogers, *Biohazard*, Knopf, New York, 1977; J. Goodfield, *Playing God*, Hutchinson, London, 1977; N. Wade, *The Ultimate Experiment: Man-Made Evolution*, Walker & Co., New York, 1977; J. Lear, *Recombinant DNA: The Untold Story*, Crown Publishers, New York, 1979; R. Hutton, *Bio-Revolution: DNA and the Ethics of Man-Made Life*, New American Library, New York, 1978. In addition, material from the MIT Archives Recombinant DNA Collection was of great importance.

3. Indeed, between 1955 and 1961, some 10–30 million US children received polio immunisation infected by live SV40, and they have not shown higher subsequent rates of malignancy appearing. See S. Krimsky, 'Regulating Recombinant DNA Research', in D. Nelkin (ed.), *Controversy: Politics of Technical Decisions*, Sage, Beverly Hills and London, 1979, pp.227–53; M. Rogers, 'The Pandora's Box Congress',

Rolling Stone, Vol. 189, 19 June 1975, p.36 ff.

4. Interview with Paul Berg by R. Goodall, 17 May 1975, MIT Archives.
5. See interview with John Tooze by C. Weiner, 26 March 1976, MIT Archives.
6. For reference to this and many of the following points, see 'Chronology', *Finding Aid*, Recombinant DNA Collection, MIT Archives.
7. Berg had been reluctant to address such wider issues, preferring to see the issue as one of health hazard only. Interview with Berg, *op. cit.*
8. See 'Chronology', *Finding Aid, op. cit.*, and N. Wade, 'Microbiology: Hazardous Profession Faces New Uncertainties', *Science*, Vol. 182, 9 November 1973, p.567.
9. See Chapter 1.
10. See interview with Berg, *op. cit.* The conference proceedings have been published. See A. Hellman, M.N. Oxman and R. Pollack (eds), *Biohazards in Biological Research*, Cold Spring Harbor Laboratory, New York, 1973. A report of the work of A. Lewis is included.
11. Some people have argued that recombinant DNA should never have been singled out for special treatment, to face constraints greater than elsewhere. The corollary of this would be that all hazardous activity should be treated as were the conjectured hazards associated with genetic manipulation. No other activity had dramatic calls for caution by those involved, except nuclear energy.
12. See N. Wade, 'Microbiology: Hazardous Profession Faces New Uncertainties', *op. cit.*, p.566. Training in microbiological techniques was, on a number of occasions, to be recommended as a precursor to carrying out genetic manipulation.
13. J.D. Watson in discussion in Hellman, Oxman and Pollack (eds), *op. cit.*, p.351. He went on to criticise the National Cancer Institute for failing to live up to its moral if not legal responsibilities.
14. Plasmid Stanley Cohen 101 or pSC101. See Chapter 1. See also S.N. Cohen, 'The Manipulation of Genes'. *Scientific American*, July 1975, pp.24–33.
15. See S.N. Cohen *et al.*, 'Construction of Biologically Functional Bacterial Plasmids In Vitro', *Proceedings of the National Academy of Sciences*, Vol. 70, No. 11, November 1973, pp.3240–44.
16. See Grobstein, *op. cit.*, p.18; M. Singer, 'Research with Recombinant DNA', *Academy Forum*, National Academy of Sciences, Washington DC, 1977; M. Singer, 'Re-examination of Basic Assumptions: Chairman's Introduction', in J. Morgan and W.J. Whelan (eds), *Recombinant DNA and Genetic Experimentation*, Pergamon Press, Oxford, New York, 1979, pp.185–6. See also M. Rogers, *Biohazard, op. cit.*; Goodfield, *op. cit.*; N. Wade, *The Ultimate Experiment, op. cit.*
17. See M. Singer and D. Söll, 'Guidelines for DNA Hybrid Molecules', *Science*, Vol. 181, 21 September 1973, p.1114.
18. Essentially covering variations on methods discussed in Chapter 1. The letter also called for some consideration of the then current large-scale preparations of animal viruses.

19. See M. Singer, letter to H.L. Kornberg, 6 June 1974, MIT Archives. Note that Kornberg at the time of the letter was a member of the Ashby Committee which was to examine the issues in terms of the UK institutional response. See Chapter 4.

20. Interview with Berg, *op. cit.*

21 See J.D. Watson, 'Why the Berg Letter was Written', in Morgan and Whelan (eds), *op. cit.*, p.190. In attendance were P. Berg, J. Watson, D. Baltimore, S. Weissman, D. Nathans, R. Roblin, N. Zinder and H. Lewis.

22. He had also proposed writing a report at the National Science Foundation Human Cell Biology Panel the previous month, although those in attendance declined to become personally involved, except for Norton Zinder.

23. In retrospect, Watson has regretted the actions they were to take. In 1979 he was to say that more expertise should have been brought together from fields involving infectious diseases. He acknowledged that in 1974 they intended to bring the expertise together later, with the microbiologists. See Watson, *op. cit.* Many of Watson's misgivings arose out of the dramatic events following their actions five years earlier. This, however, is a general problem of hindsight, and in decision-making generally. Decisions must be located within the context of their time. Others regretted the politicised response as well as Watson, and in 1977 they considered publishing a letter to neutralise the consequences of their earlier actions, in the light of new knowledge. See J.D. Watson and J. Tooze, *The DNA Story: A Documentary History of Gene Cloning*, W.H. Freeman & Co., San Francisco, 1981, pp.251–61. Those involved included Berg, Cohen, Zinder and Watson.

24. See P. Berg. Letter to H.L. Kornberg, 18 June 1974, MIT Archives. Quoted by C. Weiner in 'Historical Perspectives on the Recombinant DNA Controversy', in Morgan and Whelan (eds), *op. cit.*, pp.282–3. Weiner was responsible for setting up the archive at MIT.

25. Many accounts have just repeated the order of events as they occurred. The conference was held after the letter was published, implying, as the letter called for the conference, that the conference was a result.

26. See interview with Berg, *op. cit.* See also 'Rough Chronology of Drafts of NAS Committee Statement', handwritten by R. Roblin in MIT Archives. Note also that two of the earlier drafts, one by Roblin and one by Berg, included a paragraph on the potential use of the new techniques for biological warfare. See below.

27. See Weiner, *op. cit.*, p.283 and 'Chronology', *Finding Aid, op. cit.*

28. L. Crawford *et al.*, letter to J. Kendrew, 7 June 1974, MIT Archives. See also interview with Tooze, *op. cit.*, and see Weiner, *op. cit.*, p.283.

29. P. Berg, D. Baltimore, H. Boyer, S. Cohen, R. Davis, D. Hogness, D. Nathans, R. Roblin, J. Watson, S. Weissman, N. Zinder, 'Potential Biohazards of Recombinant DNA Molecules', *Science*, Vol. 185, 26 July 1974, p.303 and *Proceedings of the National Academy of Sciences*, Vol. 71, No. 7, July 1974, pp.2593–94. The British journal *Nature*, however, chose to use the more sensational heading of 'NAS Ban on

Plasmid Engineering' in *Nature*, Vol. 250, 19 July 1974, p.174. Further, *Nature*, in missing three words out at the start of a very important paragraph, inferred that the concern was over all work with replicating bacterial plasmids. At best it was not clear. Finally, the *Nature* version omitted the final paragraph for editorial purposes. This had emphasised the awareness of Berg and company of what they were asking researchers to do, and it emphasised the personal appeal. Its omission gave the impression of an edict, which they had tried to avoid. Some of the initial British response not surprisingly was not very favourable. E.S. Anderson wrote to *Nature*, saying that the letter addressed well understood work, and that he wished 'it had been presented less pompously'! It was not until September that Berg found out about the *Nature* version, when about to appear with Anderson on a television programme. When Anderson heard the real intent he apologised on the air. Of those who met at MIT only H. Lewis did not sign, but others did who were not there, despite the view of Berg that he wanted to avoid a petition.

30. The final draft amended the request for a 'meeting' to an 'international meeting', as a result of the growing awareness of the global nature of the research.

31. A fact on which a number of commentators have noted. See, for example, N. Wade, 'Genetic Manipulation: Temporary Embargo Proposed on Research', *Science*, Vol. 185, 26 July 1974, pp.332–34; Krimsky, *op. cit.*, p.231; S.P. Stich, 'The Recombinant DNA Debate: Some Philosophical Considerations', in D.A. Jackson and S.P. Stich (eds), *The Recombinant DNA Debate*, Prentice-Hall, Englewood Cliffs, NJ, 1979, p.183; C. Grobstein, *op. cit.*, p.21. Not all comments are specific to the Berg letter, but quite often reference is to the overall uniqueness of the debate.

32. Berg felt the press headlines were abominable, although the content was all right. See interview with Berg, *op. cit.*

33. Watson, *op. cit.*, p.191.

34. In science the incentive to publish cannot be underestimated, and the NIH examined journals published during and after the deferral. See NIH, *Environmental Impact Statement on NIH Guidelines for Research Involving Recombinant DNA Molecules*, Part One, US Department of Health, Education and Welfare, Bethesda, Maryland, 1977, p.17.

35. See N. Wade, 'Genetic Manipulation: Temporary Embargo Proposed on Research', *op. cit.*, p.332.

36. K. Hasunama, Letter to P. Berg, 15 February 1975, and R. Curtiss, 'Memorandum', both in MIT Archives. Curtiss, who was to become quite influential over the issues, also commented on how slow to act had been the nuclear scientists in addressing the problems they had unearthed. Curtiss, indeed, became something of a specialist in long, typed, single-spaced documents!

37. See C. Norman, 'NIH Backing for NAS Ban', *Nature*, Vol. 250, 26 July 1974, p.278. Note *Nature* had published the letter a week earlier than in the US and this report appeared on the day *Science* published the Berg

appeal. R.S. Stone's successor as Director, NIH, D.S. Fredrickson, would play a more important overall role.

38. The committee was established under Section 301 of the Public Health Service Act (42 USC 241) which mandates the Secretary of Health, Education and Welfare (the NIH is a section of this department) to 'conduct research, investigations, experiments, demonstrations and studies relating to the causes, diagnosis, treatment, control and prevention of physical diseases and impairments of man'. See *Charter: Recombinant DNA Molecule Program Advisory Committee*, Department of Health, Education and Welfare, Washington DC, a copy of which is in the MIT Archives.

39. *Charter*, RAC, *idem*.

40. S.N. Cohen in discussion at a meeting at Wye College, Kent, in Morgan and Whelan (eds), *op. cit.*, p.296. This international meeting is discussed elsewhere in this book.

41. See A. Ryan, *The Philosophy of the Social Sciences*, Macmillan, London, 1970, for an analysis of the philosophical issues as they apply to both science and social science.

42. In attendance were: P. Berg, M. Singer, D. Baltimore, N. Zinder, S. Weissman, R. Roblin, H. Lewis, R. Novick, W. Gartland, A. Shatkin and D. Brown. See H. Lewis, 'Biohazard Conference Organising Committee', notes, 17 October 1974, MIT Archives.

43. The eminent British scientist Sydney Brenner and Niels Jerne, the chairman of the EMBO Council.

44. See interview with Berg, *op. cit.*

45. See Lewis, *op. cit.* The list of participants is available in the MIT Archives.

46. For the importance of bacteria, plasmids and viruses in recombinant DNA technology, see Chapter 1. Viruses as a focus of study also reflected the interest in their links with tumours. Focus on the bacterium, *E. coli*, would reflect its experimental importance and its existence in humans.

47. International organisations are discussed in some depth in Chapter 5. The IAMS subsequently had declining impact in the issue area as better placed organisations became involved.

48. The meeting was organised by the Gottlieb Duttweiler Institute for Economic and Social Studies, the Swiss Society for Cell and Molecular Biology and Forum Davos. See *Applications and Limitations of Genetic Engineering: The Ethical Implications*, Proceedings: Gottlieb Duttweiler Institute, Switzerland, 1974. See also R.M. Croose Parry, 'The Promethean Situation: A Report of the Davos Conference', *Futures*, April 1975, pp.169–73.

49. For example, excessive emphasis on the achievements of Swiss science and problems of research funds. In many ways, the conference was pitched at issues which were too broad concerning science policy in general.

50. See H. Wheeler, 'The Challenge of Davos', in the Proceedings: Gottlieb Duttweiler Institute, *op. cit.*

51. *Report of the Working Party on the Experimental Manipulation of the Genetic Composition of Micro-organisms*, HMSO, London, Cmnd 5880, January 1975, See Chapter 4.
52. A number of useful sources provide insights into the content, but also into the mood of this conference. In particular, see M. Rogers, 'The Pandora's Box Congress', *op. cit.*, and *Biohazard, op. cit.* His article for *Rolling Stone* is acknowledged as one of the best summaries of the meeting. See also G. Chedd, 'Genetic Engineers Discuss Our Future', *New Scientist*, 6 March 1975, p.547; N. Wade, 'Genetics: Conference Sets Strict Controls to Replace Moratorium', *Science*, Vol. 187, 14 March 1975, pp.931–3; C. Norman, 'Berg Conference Favours Use of Weak Strains', *Nature*, Vol. 254, 6 March 1975, p.6. See also Krimsky, *op. cit.* Much material is also available at the MIT Archives regarding this conference and its preparation, including a tape recording of the proceedings.
53. See Chedd, *op. cit.*
54. Krimsky, *op. cit.*, p.233.
55. Quoted in N. Wade, 'Genetics: Conference Sets Strict Controls to Replace Moratorium', *op. cit.*, p.933.
56. N. Wade, ibid., p.932.
57. Ibid., p.933; and C. Norman, 'Berg Conference Favours Use of Weak Strains', *op. cit.*, p.7.
58. See N. Wade, 'Genetics: Conference Sets Strict Controls to Replace Moratorium', *op. cit.*, p.933 and M. Rogers, *Biohazard, op. cit.*, p.73.
59. N. Wade, 'Genetics: Conference Sets Strict Controls to Replace Moratorium', *op. cit.*, p.934.
60. They would, however, involve the use of genes specifying toxins such as botulinus. Ironically, in later years it was argued that many hazardous pathogens could be treated more safely than in their natural form if implanted in a disabled strain of *E. coli* or other bacterium.
61. See C. Norman, 'Berg Conference Favours Use of Weak Strains'. *op. cit.*, p.7.
62. A. Capron, quoted in M. Rogers, *Biohazard, op. cit.*, p.78.
63. Quoted in Rogers, ibid., p.83. This was modified by the time the final statement was published, but the intent remained.
64. Handwritten note from M. Singer to P. Berg at Asilomar. Copy in MIT Archives.
65. See P. Berg, D. Baltimore, S. Brenner, R. Roblin, M. Singer, 'Asilomar Conference on DNA Recombinant Molecules', *Nature*, Vol. 255, 5 June 1975, pp.442–3, *Science*, Vol. 188, 6 June 1975, pp.991–4 and *Proceedings of the National Academy of Sciences*, Vol. 72, No. 6, June 1975, pp.1981–84.
66. Weiner, *op. cit.*, p.287. Sixteen members of the press attended, both from science and daily newspapers.
67. The letter is contained in the MIT Archives. See also RAC, Minutes of Meeting, 28 February 1975, MIT Archives. Ten members of the press attended.
68. For many Eastern European and Soviet scientists, this seems to have

been their first knowledge of the techniques at all, according to Kaplan. See interview with M. Kaplan by C. Weiner, 8 March 1976, MIT Archives.

69. O. Maaløe, Chairman, *Scientific and Ethical Questions Involved in the Problem of a Moratorium on Certain Biological Research*, Proceedings of an Informal Session, 31 August 1974, MIT Archives.

70. M. Kaplan *et al., In Vitro Recombination (Genetic Engineering). Special Report to the Plenary Session of the 26th Pugwash Congress*, August 1976. Copy in MIT Archives.

71. Draft of the Berg letter, MIT Archives.

72. Watson, *op. cit.*, p.191. The Fort Detrick laboratory was involved in United States research into biological weapons.

73. P. Pringle and J. Spigelman, *The Nuclear Barons*, Sphere Books, London, Holt, Rinehart and Winston, New York, 1981, p.414. Paracelsus was a sixteenth-century Swiss physician and alchemist who arrogantly proclaimed himself ahead of his time and right where others were wrong.

CHAPTER 3

1. See S.N. Cohen *et al., Report to COGENE from the Working Group on Recombinant DNA Guidelines*, Committee on Genetic Experimentation, International Council of Scientific Unions, Miami, 1980, pp.34–36. See Chapter 5 for descriptions of COGENE and ICSU.

2. 'Guidelines for Research Involving Recombinant DNA Molecules', *Federal Register*, Vol. 41, No. 131, July 1976, pp.27911–22.

3. See A. Zander, 'The Discussion of Recombinant DNA at the University of Michigan', in D.A. Jackson and S.P. Stich (eds), *The Recombinant DNA Debate*, Prentice-Hall, Englewood Cliffs, NJ, 1979, p.5. Three members of the first Michigan committee had been at Asilomar II.

4. See C. Norman, 'Science Vs the Public', *Nature*, Vol. 262, 15 July 1976, pp.163–5. See also 'Cambridge (Mass.) Blocks the Genetic Engineers', *New Scientist*, 15 July 1976, p.115. Vellucci banned P3 and P4 work. The proposals had been for a P3 laboratory, see below.

5. For the views of Friends of the Earth, as expressed by its 'Committee for Genetics', see F.R. Simring, 'On the Dangers of Genetic Meddling', *Science*, Vol. 192, 4 June 1976, p.940. Overall, though, as the recombinant DNA debate unfolded nationally, there were some internal divisions. See interview with P. Lippe by A. Seidman, 13 January 1978, MIT Archives. Lippe was the Washington FoE representative most active in the issue.

6. See 'Guidelines for the Use of Recombinant DNA Molecule Technology in the City of Cambridge', January 1977, reprinted in C. Grobstein, *A Double Image of the Double Helix: The Recombinant DNA Debate*, W.H. Freeman & Co., San Francisco, 1979, pp.152–7. Also in *Bulletin of the Atomic Scientists*, Vol. 33, May 1977, pp.22–6.

7. Respectively, the University of California at San Diego, the University of Wisconsin and the University at Indiana. See N. Wade, 'Gene-Splicing: At Grass-Roots Level a Hundred Flowers Bloom', *Science*, Vol. 195, 11 February 1977, pp.558–60.
8. Epidemiological studies for example. See RAC, Minutes of Meeting, 28 February 1975, MIT Archives.
9. 'Chronology', *Finding Aid*, Recombinant DNA Collection, MIT Archives. Activities such as Senate hearings and legislation are considered below and in Chapter 6.
10. Curiously, the newsletter was initially to be distributed to those who attended Asilomar II, and after two years only to those who contributed to it at least once a year. In attendance at the meeting were representatives from the European Science Foundation (R. Litman), the NAS, the Energy Research and Development Administration and NIAID (W. Rowe). See RAC, Minutes of Meeting, 12–13 May 1975, MIT Archives.
11. See RAC, Minutes of Meeting, 18–19 July 1975, MIT Archives and N. Wade, 'Recombinant DNA: NIH Group Stirs Storm by Drafting Laxer Rules', *Science*, Vol. 190, 21 November 1975, p.767. See also NIH, *Environmental Impact Statement on NIH Guidelines for Research Involving Recombinant DNA Molecules*, Part One, US Department of Health, Education and Welfare. Bethesda, Maryland, 1977, p.20; and *Genetic Engineering, Human Genetics and Cell Biology: Evolution of Technical Issues, RNA Recombinant Molecule Research (Supplement Report II)*, Subcommittee on Science, Research and Technology of the Committee on Science and Technology. US House of Representatives, Ninety-Fourth Congress, 2nd Session, Library of Congress, Serial KKK, Washington DC, 1976, p.21.
12. D. Stetten, letter to M.H. Edgell, University of North Carolina, 9 September 1975, MIT Archives.
13. See R. Goldstein, *et al.*, petition sent to D. Stetten, NIH, 27 August 1975, MIT Archives. Most of the letter has been reproduced in 'DNA Committee has its Critics', *Nature*, Vol. 257, 23 October 1975, p.637.
14. R. Curtiss, letter to D. Stetten, NIH, 13 August 1975, MIT Archives. See also C. Norman, 'Genetic Manipulation: Recommendations Drafted', *Nature*, Vol. 258, 18 December 1975, pp.561–4.
15. See J. King, 'A Science for the People', *New Scientist*, 16 June 1977, pp.634–6.
16. J. King, ibid., pp.634–5.
17. See D. Hogness, Letter to L. Jacobs, NIH, 10 November 1975, MIT Archives.
18. See C. Norman, 'Genetic Manipulation: Recommendations Drafted', *op. cit.*, p.562; and N. Wade, 'Recombinant DNA: NIH Sets Strict Rules to Launch New Technology', *Science*, Vol. 190, 19 December 1975, pp.1175–79.
19. See RAC, Minutes of Meeting, 4–5 December 1975, MIT Archives. See also C. Norman, 'Genetic Manipulation: Recommendations Drafted', *op. cit.*, pp.562–3.

20. An unnamed committee member quoted in C. Norman, ibid., p.562. See also N. Wade, 'Recombinant DNA: NIH Sets Strict Rules to Launch New Technology', *op. cit.*, p.1176.
21. Wade, ibid., p.1178.
22. See RAC, Minutes of Meeting, 4–5 December 1975, *op. cit.*, and Wade, 'Recombinant DNA: NIH Sets Strict Rules to Launch New Technology', *op. cit.*
23. See D.S. Fredrickson, 'Decision of the Director, NIH, to Release Guidelines for Research on Recombinant DNA Molecules, June 1976', *Federal Register*, Vol. 41, No. 131, July 1976, pp.27901–11.
24. See *Genetic Engineering, Human Genetics and Cell Biology: Evolution of Technical Issues, op. cit.*, p.23 and Appendix 6, pp.103–8. Notified were 17 groups such as FAS, FoE, the League of Women Voters, Consumer Federation of America, Centre for Law and Social Policy, Environmental Defense Fund, etc. See also Transcript of the Meeting of *Ad Hoc* Advisory Committee to the Director, NIH, 9–10 February 1976, MIT Archives.
25. M. Rogers, *Biohazard*, Knopf, New York, 1977, p.178. Included were a Chief Judge, a hospital medical director, the Provost of MIT, the Director of the Institute of Society, Ethics and the Life Sciences, a lawyer and the President of the National Consumers' League.
26. R. Sinsheimer, letter to D.S. Fredrickson, NIH, 5 February 1976, MIT Archives.
27. N. Wade also argued that human error would be likely. He noted that of the 5000 cases of laboratory-acquired infections over 30 years to that date, a third occurred in laboratories with special containment facilities. Even in the highest containment of the US Army's biological warfare laboratories, there were over 423 cases of infection and three deaths over 25 years. He emphasised these occurred with *known* hazardous organisms. N. Wade, 'Go-Ahead for Recombinant DNA', *New Scientist*, 18/25 December 1975, p.684.
28. See Chapter 1.
29. Sinsheimer, *op. cit.*, p.3.
30. R. Sinsheimer, letter to D.S. Fredrickson, NIH, 12 February 1976, MIT Archives.
31. See E. Chargaff, letter to D.S. Fredrickson, NIH, 8 February 1976, MIT Archives, and E. Chargaff, 'On the Dangers of Genetic Meddling', *Science*, Vol. 192, June 1976, pp.938–9.
32. See N. Wade, *The Ultimate Experiment: Man-Made Evolution*, Walker & Co., New York, 1977, pp.110–11.
33. See P. Berg, letter to D.S. Fredrickson, NIH, 17 December 1976; and M. Singer, letter to D.S. Fredrickson, NIH 13 February 1976. Goldstein, co-organiser of the Cold Spring Harbor petition, was making a comparison with nuclear energy. See R. Goldstein, letter to D.S. Fredrickson, NIH, 13 February 1976. These and many other letters, on the above proposed guidelines and meetings, are contained in the MIT Archives.
34. D.S. Fredrickson, 'Selected Issues for Committee Review', 19 March 1976, MIT Archives.

35. See N. Wade, 'Recombinant DNA: The Last Look Before the Leap', *Science*, Vol. 192, 16 April 1976, pp.236–8. Wade regretted that minority views such as Sinsheimer's, despite their weight, were not discussed.

36. See P. Berg, letter to D.S. Fredrickson, NIH, 6 April 1976, MIT Archives and J.F. Kelly, letter to D.S. Fredrickson, NIH, 14 April 1976, MIT Archives.

37. The number of respondents was 322 who replied to a request in the FAS monthly publication. See FAS, 'Results of DNA Straw Poll from April Public Interest Report', MIT Archives. See also letters to D.S. Fredrickson, NIH, from the following: B. Trumball, 15 April 1976; R.N.L. Andrews, 9 June 1976; L. Salzman (Friends of the Earth), 17 May 1976; A. Schwartz, S. Wright, M. Ross, R.P. Weeks, M. Heirich, 21 April 1976; all in MIT Archives.

38. See N. Wade, *The Ultimate Experiment, op. cit.*, p.102. By this time the guidelines were published.

39. See G. Wald *et al.*, Petition, June 1976, MIT Archives.

40. See 'Guidelines ...', *op. cit.*; and 'Decision of the Director, National Institutes of Health, to Release Guidelines for Research on Recombinant DNA Molecules', *Federal Register*, Vol. 41, No. 131, July 1976, pp.27902–11. See also C. Norman, 'Genetic Manipulation: Guidelines Issued', *Nature*, Vol. 262, 1 July 1976, p.2.

41. 'Decision of the Director ...', *op. cit.*, p.27905. Such work was to be allowed, but with stringent precautions.

42. Classes 3, 4 and 5 of the Department of Health, Education and Welfare, *Classification of Etiological Agents on the Basis of Hazard*, Public Health Service, Center for Disease Control, Office of Biosafety, Atlanta, Georgia, 1974.

43. Some exceptions could be considered on the last point if 'social benefits' would be great and extra containment precautions were taken. See 'Guidelines ...', *op. cit.*, pp.27914–17.

44. See, for example, C. Norman, 'Genetic Manipulation: Guidelines Issued', *op. cit.*, and C. Grobstein, 'The Recombinant DNA Debate', in *Recombinant DNA: Readings from Scientific American*, W.H. Freeman & Co., San Francisco, 1978, p.141 and reprinted in C. Grobstein, *op. cit.*, pp.32–3.

45. See 'Guidelines ...', *op. cit.*, p.27912.

46. Department of State, Telegram, June 1976, in J.D. Watson and J. Tooze, *The DNA Story: A Documentary History of Gene Cloning*, W.H. Freeman & Co., San Francisco, 1981, p.81.

47. See 'Decision of the Director ...', *op. cit.*, pp.27905–6.

48. W.N. Hubbard, letter to D.S. Fredrickson, NIH, 16 July 1976, MIT Archives.

49. C.W. Pettinga, letter to D.S. Fredrickson, NIH, 4 June 1976, MIT Archives. Pettinga sent this letter two days after attending the meeting with Fredrickson mentioned above.

50. *Genetic Engineering, Human Genetics and Cell Biology: Evolution of Technical Issues, op. cit.*, p.51.

51. See W. J. Gartland, Memorandum, 18 June 1976, Appendix C. NIH, *Environmental Impact Statement, op. cit.*, Part Two.
52. See W.J. Gartland, ibid.
53. See 'Decision of the Director ...', *op. cit.*, pp.27910–11.
54. US National Environmental Policy Act, 1969, quoted by R.M. Hartzman, letter to the Office of General Counsel, Department of Health, Education and Welfare, 16 March 1977, MIT Archives. Hartzman was a lawyer retained by Friends of the Earth, who were pressing for an Environmental Impact Statement.
55. 'Decision of the Director ...', *op. cit.*, p.27906.
56. 'Recombinant DNA Research Guidelines: Draft Environmental Impact Statement', *Federal Register*, Vol. 41, No. 176, 9 September 1976, pp.38425–83.
57. NIH, *Environmental Impact Statement, op. cit.*, Parts One and Two, October 1977.
58. 'Recombinant DNA Research Guidelines: Draft EIS', *op. cit.*, p.38434.
59. The complete list can be found in NIH, *Environmental Impact Statement, op. cit.*, Part Two, Appendix I.
60. NIH, ibid., Appendix J. The following laws were examined in detail: The Occupational Safety and Health Act, 1970 (Public Law 91–596); The Toxic Substances Control Act (Public Law 94–469); The Hazardous Materials Transportation Act (Public Law 93–633); Section 361 of the Public Health Service Act (42 USC Sec. 264).
61. Interim Report of the Federal Interagency Committee, in NIH, *Environmental Impact Statement, op. cit.*, p.10.
62. L.J. Lefkowitz, Attorney General of the State of New York, Comments on the Guidelines and Draft EIS, in NIH, *Environmental Impact Statement, op. cit.*, Part Two, Appendix K. Indeed all the comments were collated in this appendix.
63. NIH, *Environmental Impact Statement, op. cit.*, Part One, p.118.
64. See P. David, 'Rifkin's Regulatory Revivalism Runs Riot', *Nature*, Vol. 305, 29 September 1983, p.349. As the NIH is not a regulatory agency the public do not have the same right to challenge its advice in the courts as they could regarding a regulatory agency.
65. Formerly, the People's Business Commission.
66. See C. Joyce, 'Court Blocks Field Trial of Spliced Genes', *New Scientist*, 24 May 1984, p.7 and S. Budiansky, 'Rifkin Strikes at Corn This Time', *Nature*, Vol. 310, 5 July 1984, p.3.
67. See S. Budiansky, 'Rifkin Battle Lost, War Undecided', *Nature*, Vol. 314, 7 March 1985, p.6. This was not to be the end of the problems facing Advanced Genetic Sciences. In February 1986, despite the company having approval from the EPA and the California Department of Food and Agriculture, local politicians in Monterey halted a similar experiment involving environmental release of a bacterium which might restrict frost damage to strawberry crops. US researchers in both industry and academia would once again face the local level of expressed uncertainties and decision processes.
68. See Department of Health, Education and Welfare, NIH,

'Recombinant DNA Research; Physical Containment Recommendations for Large-Scale Uses of Organisms Containing Recombinant DNA Molecules', *Federal Register*, 11 April 1980, pp. 24968–71. For a discussion of the safety issue with respect to 'bioreactors', see D.F. Liberman, R. Fink and F. Schaefer, 'Biosafety and Bioreactors', *Biotech 84: Europe*, Online Publications, Pinner, 1984, pp. 103–14.

69. See *Impacts of Applied Genetics: Micro-organisms, Plants and Animals*, Office of Technology Assessment, US Congress, April 1981, p.218.
70. See S. Budiansky, 'Regulation Issue is Resurrected by EPA', *Nature*, Vol. 304, 18 August 1983, p.572.
71. On this point see S. Wright, 'Molecular Politics in Great Britain and the United States: The Development of Policy for Recombinant DNA Technology', *Southern California Law Review*, Vol. 51, No. 6, September 1978, pp.1383–34.

CHAPTER 4

1. See S.N. Cohen *et al.*, *Report to COGENE from the Working Group on Recombinant DNA Guidelines*, Committee on Genetic Experimentation, International Council of Scientific Unions, Miami, 1980, p.37. By comparison, Japan had some 35 laboratories involved, Federal Republic of Germany 10–20, Canada 10–15, France 12, Australia 16, Switzerland 18 and all other states were in single figures. The estimates are derived from responses to a questionnaire organised by COGENE.
2. See S. Wright, 'Molecular Politics in Great Britain and the United States: The Development of Policy for Recombinant DNA Technology', *Southern California Law Review*, Vol. 51, No. 6, September 1978, pp.1383–434.
3. *Report of the Committee of Inquiry into the Smallpox Outbreak in London in March/April 1973*, HMSO, London, Cmnd 5626, 1974.
4. See B.A. Turner, *Man-Made Disasters*, Wykeham Publications, London. 1978, pp.109–17.
5. Or a delay over generations where the host passed the information through the strain, or related strains.
6. *Report of the Working Party on the Laboratory Use of Dangerous Pathogens*, HMSO, London, Cmnd 6054, 1975. Subsequently referred to as the Godber Report, after its chairman, Sir George Godber. Godber began his investigation before the Ashby working party, but published afterwards.
7. See Wright, *op. cit.*, pp.1393–94, and Chapter 2. See also 'Genetics: Conference Sets Strict Guidelines to Replace Moratorium', *Science*, Vol. 187, 14 March 1975, p.932. Note that the MRC advises the University Grants Committee on, among other things, laboratory funding.
8. See E. Ashby, *ABRC Working Party on Genetic Manipulation of Micro-*

organisms: Note from the Chairman, ABRC (GE) 74/3, 26 August 1974, MIT Archives.

9. *Report of the Working Party on the Experimental Manipulation of the Genetic Composition of Micro-organisms*, HMSO, London, Cmnd 5880, January 1975, p.3. Subsequently referred to as the Ashby Report, after its chairman. See also E. Ashby, letter to P. Berg, 5 November 1974, MIT Archives. In the letter Ashby expressed his intentions of making the Report a discussion document.

10. Ashby Report, *op. cit.*, p.2.

11. The techniques of achieving recombinant DNA molecules were not discussed in any depth, but at that time referred to methods similar to those using restriction enzymes outlined in Chapter 1.

12. Including bacteriology, cancer research, molecular biology in general, virology, biochemistry. The representatives were scientists with similar interests and could perhaps be classed as a distinct sociological group. See Ashby Report, *op. cit.*, Appendix 2, 'List of Expert Witnesses'.

13. Ashby Report, *op. cit.*, pp.6–7.

14. See B. Dixon, 'Not Good Enough', *New Scientist*, 23 January 1975, p.186. Dixon applauds the motives of the Ashby group and the effect of stimulating discussion.

15. Ashby Report, *op. cit.*, p.10.

16. *Report of the Working Party on the Practice of Genetic Manipulation*, HMSO, London, Cmnd 6600, August 1976, p.11. Subsequently referred to as the Williams Report, after its chairman, Sir Robert Williams. See below.

17. Ashby Report, *op. cit.*, p.9.

18. In the US, for example, the National Institutes of Health (NIH) Environmental Impact Statement, of October 1977, recognised the possibilities of 'deliberate misuse', but argued that as far as weapons were concerned, the application of recombinant DNA techniques would be outlawed under the Biological Weapons Convention. Discussion was limited in effect to this observation. See NIH, *Environmental Impact Statement on NIH Guidelines for Research Involving Recombinant DNA Molecules*, Part One, US Department of Health, Education and Welfare, Bethesda, Maryland, 1977, pp.38–9.

19. See Dixon, *op. cit.*, p.186.

20. 'Amber Light for Genetic Manipulation', editorial, *Nature*, vol. 253, 31 January 1975, p.295.

21. Ashby Report, *op. cit.*, pp.12–13.

22. E. Yoxen, 'Regulating the Exploitation of Recombinant Genetics', in R. Johnston and P. Gummett (eds), *Directing Technology*, Croom Helm, London, 1979, pp.228–9. Yoxen refers to issues such as the medical policy or sort of technology which might be needed in the UK and the question of socially acceptable risk.

23. DES, 'Genetic Manipulation of Micro-organisms', press notice issued on 6 August 1975, MIT Archives. As press coverage to date had been thin, the notice was more widely distributed to interested parties.

24. F. Mulley, quoted in DES, press notice, ibid., p.1.

25. Williams Report, *op. cit.*, p.3.
26. 'Forever Amber on Manipulating DNA Molecules?', editorial, *Nature*, Vol. 256, 17 July 1975, p.155. The conference had been organised under the auspices of the Research Councils and the Department of Health and Social Security.
27. The Birmingham smallpox outbreak, see below.
28. See *Second Meeting of the EMBO Standing Advisory Committee on Recombinant DNA, held at London on 18/19 September 1976: Report and Recommendations*, EMBO, Heidelberg, 1976. See also J. Tooze, 'Genetic Engineering in Europe', *New Scientist*, 10 March 1977, pp.592-4.
29. Williams Report, *op. cit.*, p.8
30. 'Genetic Guidelines: Handle with Care', editorial, *Nature*, Vol. 263, 2 September 1976, p.1. See also E. Lawrence, 'Genetic Manipulation: Guidelines Out', in the same edition, pp.4-5.
31. See Tooze, *op. cit.*, p.592.
32. See F.R. Simring, letter to D.S. Fredrickson, NIH, 19 June 1978, MIT Archives. Simring was the Executive Director of the Coalition for Responsible Genetic Research. She expounded the virtues of centralised advisory groups such as that of the UK.
33. See C. Sherwell, 'Heading for Harmony?', *Nature*, Vol. 266, 3 March 1977, pp.2-4.
34. See Williams Report, *op. cit.*, p.13.
35. See R. Freedman, 'Gene Manipulation: A New Climate', *New Scientist*, 27 July 1978, pp.268-9. See also 'Memorandum submitted by the Genetic Engineering Group of the British Society for Social Responsibility in Science', in *Second Report from the Select Committee on Science and Technology, Recombinant DNA Research–Interim Report*, HMSO, London, HC 355, 1979, p.206. Both of these references cite the uniqueness of GMAG.
36. See GMAG, *First Report of the Genetic Manipulation Advisory Group*, HMSO, London, Cmnd 7215, 1978, pp.vii-viii. In addition, meetings were attended by a number of assessors from various government departments.
37. See 'Blind Man's Buff at GMAG', *Nature*, Vol. 276, 14 December 1978, p.657. For an example of Ravetz' work on risk and society, see J.R. Ravetz, 'The Political Economy of Risk', *New Scientist*, 8 September 1977, pp.26-7.
38. See *Second Report from the Select Committee on Science and Technology, op. cit.*, p.156.
39. See R. Lewin, 'GMAG Cold Shoulders AUT', *New Scientist*, 13 January 1977, p.61.
40. See S.J. Pirt, letter to the Clerk of the Subcommittee of the Select Committee on Science and Technology, in *Second Report from the Select Committee on Science and Technology, op. cit.*, Appendix 15, pp.245-6.
41. See *Second Report from the Select Committee on Science and Technology, op. cit.*, pp.154-6.

42. See Godber Report, *op. cit.*, p.10.
43. 'Unions Move in on Dangerous Organisms', *New Scientist*, 21/28 December 1978, p.915.
44. See *Report of the Investigation into the Causes of the 1978 Birmingham Smallpox Occurrence*, HMSO, London, HC 1979–80, 668, 1980. Subsequently referred to as the Shooter Report, after Professor Reginald Shooter who conducted the inquiry.
45. See *Nature*, Vol. 277, 11 January 1979, pp.75–81. The editorial, and a number of reporters examine the Shooter Report. See also E. Yoxen, *The Gene Business: Who Should Control Biotechnology?*, Pan Books, London and Sydney, 1983, pp.56–7. Some consideration was given to prosecuting ASTMS under the Official Secrets Act, but this did not occur. The report itself was eventually published.
46. 'All Safety Nets Failed, Says Shooter', *Nature*, Vol. 277, 11 January 1979, p.78.
47. See, for example, R.A. Bird, 'DPAG Needs Public Interest Representatives', letter to *Nature*, Vol. 278, 26 April 1979, p.776; R. McKie, 'Union Steps up Lab Safety Campaign', *Nature*, ibid., p.772; 'Plenty for GMAG to Do', *Nature*, Vol. 276, 2 November 1978, p.1. See also E. Yoxen, *The Gene Business, op. cit.*, p.56.
48. See A. Hay, 'Health and Safety 3 Years On', *Nature*, Vol. 270, 10 November 1977. See also Annual Reports of the HSC.
49. Williams Report, *op. cit.*, p.16.
50. See R. Lewin, 'Genetic Engineering and the Law', *New Scientist*, 28 October 1976, p.200.
51. See GMAG, *First Report, op. cit.*, pp.4–5; Lewin, 'Genetic Engineering and the Law', *op. cit.*, p.221; Hay, *op. cit.*, pp.91–2; 'Britain and the US Discuss Genetic Engineering', *New Scientist*, 18 November 1976, p.372; P. Newmark, 'UK Extends the Law to Genetic Engineering', *Nature*, Vol. 273, 15 June 1978, p.482; 'Law Catches Up with Genetic Engineering', *New Scientist*, 15 June 1978, p.732. All of these recognised that the scientific community were critical.
52. Quoted in R. Lewin, 'Genetic Engineering and the Law', *op. cit.*, and *Draft Regulation 2. Health and Safety (Genetic Manipulations) Regulations*, HSC, 1976.
53. See a published exchange of letters between M. Ashburner and J.H. Locke (HSE), *Nature*, Vol. 264, 4 November 1976, pp.2–3. See also J.A.W. McDonald (HSE) letter, *New Scientist*, 11 November 1976. McDonald criticised Lewin, who replied in the same issue, p.353.
54. See 'Britain and US Discuss Genetic Engineering', *op. cit.*, p.372.
55. The definition adopted was: 'For the purpose of these regulations, genetic manipulation shall be defined as the formation of new combinations of heritable material by the insertion of nucleic acid molecules produced, by whatever means, outside the cell, into any virus, bacterial plasmid or other vector system so as to allow their incorporation into a host organism in which they do not naturally occur but in which they are capable of continued propagation.'
56. HSE, *Health and Safety at Work: Genetic Manipulation*, HMSO, London, 1978.

57. See GMAG, *First Report, op. cit.*, pp.8–10.
58. Ibid., pp.9–10.
59. See evidence given by the HSE before the Select Committee on Science and Technology (Genetic Engineering Subcommittee) in *Second Report from the Select Committee on Science and Technology, op. cit.*, p.68.
60. Magistrates were eventually to clear Birmingham University of specific charges regarding the blame for the death of the photographer. See L. McGinty, 'An Unsavoury Outbreak', *New Scientist*, 31 July 1980, p.348. See also E. Yoxen, *The Gene Business, op. cit.*, p.57. The case was brought by ASTMS.
61. See GMAG, *First Report, op. cit.*, p.7.
62. Ibid., p.6.
63. Ibid., pp.24–6 and, for membership of the committee, pp.38–9. See also R. Lewin, 'GMAG Falls Foul of Privacy Constraints', *New Scientist*, 15 December 1977, p.683.
64. See D. Dickson, 'GMAG: Stormy Weather Ahead', *Nature*, Vol. 271, 5 January 1978, p.5. See also the evidence of J. Maddox in *Second Report from the Select Committee on Science and Technology, op. cit.*, pp.39–40.
65. GMAG, *Second Report of the Genetic Manipulation Advisory Group*, HMSO, London, Cmnd 7785, 1979, p.14.
66. In *Second Report from the Select Committee on Science and Technology, op. cit.*, p.159.
67. Ibid., p.vii.
68. See R. Lewin, 'GMAG Falls Foul of Privacy Constraints', *op. cit.*, p.683.
69. See Memorandum from the Assocation of the British Pharmaceutical Industry and Memorandum from ICI, in the *Second Report from the Select Committee on Science and Technology, op. cit.*, pp.195–205, 238–41.
70. With exceptions, such as the tragedy surrounding the Spanish poisoned cooking oil case, where the chemistry of the poison proved extremely difficult to identify, in order then to understand the biological effects.
71. GMAG has since been put under the auspices of the HSE under the new title of the 'Advisory Committee on Genetic Manipulation'. This was some five years after the ABPI memorandum. See below.
72. See GMAG, *First Report, op. cit.*, pp.18–19.
73. See D. Dickson, 'US to Increase Public Participation in Regulation of DNA Research', *Nature*, Vol. 276, 30 November 1978, p.30. Membership was to go from 14 to 20, with six public interest representatives. See also US Department of Health, Education and Welfare, 'Guidelines for Research Involving Recombinant DNA Molecules', *Federal Register*, Vol. 45, No. 20, 29 January 1980.
74. One rather ineffectual public meeting of GMAG was held on 22 December 1978. Mainly scientists attended. See E. Lawrence, 'Bacteriologists Lobby GMAG's First Public Meeting', *Nature*, Vol. 277, 4 January 1979, p.3.
75. See Wright, *op. cit.*, p.1406. See also E. Yoxen, 'Regulating the Exploitation of Recombinant Genetics', *op. cit.*, p.232.
76. See R. Lewin, 'The View of a Science Journalist', in J. Morgan and W.J.

Whelan (eds), *Recombinant DNA and Genetic Experimentation*, Pergamon Press, Oxford, New York, 1979, pp.273–6. Lewin both criticised the scientists' secrecy and commented on the lack of UK public debate, which he partly attributed to national character. He also contrasted the British Official Secrets Act with the US Freedom of Information Act.

77. 'Genetic Manipulation: New Guidelines for UK', *Nature*, Vol. 276, 9 November 1978, pp.104–8.

78. 'Manipulating MPs', *New Scientist*, 22 June 1978, p.848. See also L. McGinty, 'Grabbing the Tiger', *New Scientist*, 15 June 1978, p.730.

79. See *Second Special Report from the Select Committees on Science and Technology*, Session 1977–78, HMSO, London, HC 609, 1978 (not to be confused with the *Second Report, op. cit.*). See also 'Manipulating MPs', *op. cit.*

80. *Second Report from the Select Committee on Science and Technology, op. cit.*, p.v.

81. GMAG, the HSE, the DES, the DHSS. the Association of University Teachers, the Secretaries of State for Education and Science and Social Services, all appeared before the subcommittee. In addition, some 30 documents were also submitted from both the above groups and many other interest groups.

82. See R. McKie, 'Britain's Shadow Science Minister Believes in Experts', *Nature*, Vol. 278, 29 March 1979, p.387. The Prime Minister has, however, expressed that she was personally sorry that the House decided to disband the Select Committee. See the exchange of letters between I. Lloyd, a Conservative MP, and M. Thatcher, PM, under the title 'Why Britain Does Not Need a Minister for Science', *Nature*, Vol. 281, 27 September 1979, p.249.

83. R. Lewin, 'Genetic Engineering Under the Parliamentary Microscope', *New Scientist*, 9 August 1979, pp.430–31.

84. M.G.P. Stoker, 'Introduction and Welcome', in Morgan and Whelan (eds), *op. cit.*, p.xix.

85. See 'Royal Society President Questions Anti-Science Dogma', *New Scientist*, 7 December 1978, p.748. The other reason was fear of disaster.

86. The DES, DHSS, DoE and the Ministry of Agriculture, Fisheries and Food. The latter would find itself involved if genetic manipulation of plant DNA was proposed. It would act in consultation with GMAG adopting an HSE type role applicable this time to plants. Work with plant pathogens already required a licence by the MAFF. See 'Memorandum Submitted by the Ministry of Agriculture, Fisheries and Food (MAFF) on Behalf of the United Kingdom Agriculture Departments', in *Second Report from the Select Committee on Science and Technology, op. cit.*, pp.266–7. See also GMAG, *First Report, op. cit.*, pp. 26–9. Work involving animal DNA was covered by GMAG.

87. See Wright, *op. cit.*, pp.1400–1.

88. See *Second Report from the Select Committee on Science and Technology, op. cit.*, pp. 171–87.

89. See ibid., evidence from both the Secretaries of State for Education and Science and for Social Services.
90. See ibid., pp.94–5.
91. See P. Newmark, 'Approval for First British Virus Release Experiment', *Nature*, Vol. 320, 6 March, 1986, p.2; 'Watchdog to Bark Less Often', *Nature*, Vol. 302, 7 April 1983, p.467; and S. Yanchinski, 'Keeping the Gene Genie in the Bottle', *New Scientist*, 14 April 1983, p.69.
92. See House of Lords, Select Committee on the European Communities, *Genetic Manipulation (DNA)*, HMSO, London, HL 188, 1980. See Chapter 5.
93. Many are collected in their Second Report, *op. cit.*
94. For example, J. Ravetz and J. Maddox.

CHAPTER 5

1. See A.T. Bull *et al.*, *Biotechnology: International Trends and Perspectives*, OECD, Paris, 1982, Appendix IV.
2. S.N. Cohen *et al.*, *Report to COGENE from the Working Group on Recombinant DNA Guidelines*, Committee on Genetic Experimentation, International Council of Scientific Unions, Miami, 1980, and *Report of the Federal Interagency Committee on Recombinant DNA Research: International Activities*, US Department of Health, Education and Welfare, November 1977.
3. See interview with P. Kourilsky by C. Weiner, 20 March 1976, MIT Archives.
4. See interview with G. Bernardi by C. Weiner, 29 August 1977, MIT Archives, and Interview with P. Kourilsky, *op. cit.*
5. See J. Tooze, *Emerging Attitudes and Policies in Europe*, a review prepared for the Miles Symposium, June 1976, MIT Archives
6. See C. Sherwell, 'Heading for Harmony?', *Nature*, Vol. 266, 3 March 1977, p.2. See also minutes of the meeting of the ESF Liaison Committee for Recombinant DNA Research, 22–23 May 1978, European Science Foundation (ESF), Strasbourg, and *Report of the Federal Interagency Committee*, *op. cit.* pp.23–4.
7. By 1980, the membership was 4 experts in genetic manipulation, 4 scientists not in the field and 4 'outstanding individuals'. See Cohen *et al.*, *op. cit.* p.69.
8. See minutes of the ESF Liaison Committee, *op. cit.* See C. Conzelmann and D. Claveloux, 'Europe Fails to Agree on Biotech Rules' *New Scientist*, 10 July 1986, pp.19–20.
9. See *Guidelines for the Handling of Recombinant DNA Molecules and Animal Viruses and Cells*, Minister of Supply and Services, Canada, 1977, MR 21–1/1977. See also 'Canada: Guidelines Recommended', *Nature*, Vol. 256, 17 February 1977, p.577.
10. See Cohen *et al.*, *op. cit.*, p.68.
11. Masami Tanaka, 'Biotechnology in Japan' in *Biotech 83: Proceedings of*

the International Conference on the Commercial Applications and Implications of Biotechnology, Online Publications, Northwood, UK, 1983, p.3. Tanaka is the Director of the Bioindustry Office at the Ministry of International Trade and Industry.

12. See Y. Tazima, *Reports of Activities Displayed in Japan in Relation to Recombinant DNA Research*, prepared for an ICSU meeting, 1–2 July 1976. MIT Archives. The ICSU is discussed below.

13. See Y. Tazima, ibid., p.2. 320 scientists had been polled and of the 111 who replied, 80 per cent saw that the Berg calls as acceptable.

14. See *Splicing Life*, The President's Commission for the Study of Ethical Problems in Medicine and Biomedical and Behavioural Research, Washington DC, November 1982.

15. See *Statement on the Security of the Recombinant DNA Researches in Japan*, Adopted by the 73rd General Assembly of SCJ, 28 October 1977, MIT Archives.

16. See Cohen et al., op. cit., p.69.

17. The report by S.N. Cohen et al, and the US Interagency Committee, both referenced above, and many news items from the science news press have been of great help. In addition use has been made of the minutes of a number of meetings of international organisations, discussed below.

18. See S. Yanchinski, *Setting Genes to Work: The Industrial Era of Biotechnology*, Penguin, Harmondsworth, 1985, pp.128–9.

19. See L. Crawford et al., letter to Sir John Kendrew, 7 June 1974, MIT Archives.

20. Because not all members of the EMBC wished to be involved in the new laboratory complex, a new intergovernment structure was set up. Minutes of various meetings of the European Molecular Laboratory Council and EMBO which were made available directly from the organisation concerned, or through the MIT Archives, have been of assistance in this section.

21. Their report is held in the MIT Archives.

22. See the minutes of meetings and reports of these various bodies, and interview with John Tooze by C. Weiner, 25 March 1976, MIT Archives.

23. See *Second Meeting of the EMBO Standing Advisory Committee on Recombinant DNA, held at London on 18/19 September 1976: Report and Recommendations*, EMBO, Heidelberg, 1976. Subsequent meetings examined the drafts and finalised guidelines of other states.

24. *Second Meeting of the EMBO Standing Advisory Committee on Recombinant DNA, op. cit.*

25. ESF, *Recommendations Concerning Recombinant DNA Research*, Strasbourg, 1976. See also Report of the Federal Interagency Committee, op. cit., pp.39–41 and the annual reports of the ESF, Strasbourg.

26. A reading of the comprehensive minutes of the meetings of the ESF Liaison Committee shows it to have been very well informed and up to date. Overall trends in technical and wider issues were related to the

developments in individual countries.

27. *ESF Report 1980*, Strasbourg, 1980, p.24.
28. Interview with J. Kendrew by C. Weiner, 25 March 1976, MIT Archives.
29. See Cohen *et al., op cit.;* A.M. Skalka *et al., First Report to COGENE from the Working Group on Risk Assessment*, ISCU, July 1978; and other reports by the groups published in J. Morgan and W.J. Whelan (eds), *Recombinant DNA and Genetic Experimentation*, Pergamon Press, Oxford, New York, 1979.
30. E. Yoxen, *The Gene Business: Who Should Control Biotechnology?*, Pan Books, London and Sydney, 1983, p.59.
31. Skalka *et al., op. cit* p.iv.
32. See W. Arber *et al., Genetic Manipulation: Impact on Man and Society*, Cambridge University Press (for ICSU Press), London, New York, 1984. This was the published record of the third international symposium held to look at genetic manipulation, and comprised mainly state of the art descriptions of the science.
33. M.G.P. Stoker, 'Introduction and Welcome', in J. Morgan and W.J. Whelan (eds), *op. cit.*, p.xix. This volume represents the published proceedings of that meeting.
34. Minutes of the Second Meeting of COGENE, ICSU.
35. See R. Walgate, 'COGENE Plays Up Benefits, Plays Down Risks', *Nature*, Vol. 278, 5 April 1979, p.496. See also minutes of the Third Meeting of COGENE, ICSU.
36. 'Recombinant DNA–How Public?', *Nature*, Vol. 278, 29 March 1979, p.383.
37. 'Molecular Biology Suffering from Shock'. *Nature*, Vol. 278, 12 April 1979, p.587. See also R. Lewin, 'The View of a Science Journalist', in Morgan and Whelan (eds), *op. cit.*, pp.273–80; R. Lewin, 'Science and Politics in Genetic Engineering', *New Scientist*, 12 April 1979, pp.114–15; and 'Environmentalists Criticise Secrecy of Recombinant DNA Meeting', *Nature*, Vol. 278, 5 April 1979, p.499. US environmental groups criticised, in addition, the $200 attendance fee.
38. During discussion following Lewin's presentation at the meeting.
39. D. Haber in discussion. See J. Morgan and W.J. Whelan (eds), *op. cit.*, p.237. Haber also noted that the speakers were all known in advance to oppose guidelines and regulations.
40. Other options available to the Commission of the EC would be: *Regulations* which are binding on member states with the same strength as national laws; *Decisions* which are only binding on named parties (individuals, organisations or governments); *Recommendations* which are not binding.
41. See Sherwell, *op. cit.*, p.5.
42. See Commission of the European Communities, *Proposal for a Council Directive: Establishing Safety Measures Against the Conjectural Risks Associated with Recombinant DNA Work*, Brussels, December 1978, COM(78)664 final, p.9.
43. Ibid., p.8.

44. See the House of Lords, Select Committee on the European Communities, *Genetic Manipulation (DNA)*, HMSO, London, HL 188, 1980, p.5.

45. See Confederation of British Industry, brief submitted, in House of Lords, Select Committee, *op. cit.*, p.33.

46. See 'Memorandum of Evidence from the TUC to the Health and Safety Commission Concerning a Proposal for an EEC Directive Concerning the Conjectural Hazards of Recombinant DNA Work', in ibid., pp.35–6.

47. See J. Becker, 'DNA Research Guidelines: Dutch Get Tough', *Nature*, Vol. 290, 9 April 1981, p.439, and R. Walgate, 'European Biotechnology: EEC Proposals', *Nature*, Vol. 286, 7 August 1980, p.548.

48. See Conzelmann and Claveloux, *op. cit.* The Commission also wanted to include regulations on *in vitro* fertilisation.

49. See Commission of the European Communities, *Proposal for a Multiannual Community Programme of Research and Development in Biomolecular Engineering (indirect action 1981–1985)*, January 1980, Brussels COM(79)793 final, p.1. See also Commission of the European Communities, *A Common Policy in the Field of Science and Technology*, 1977, Brussels, COM(77)283 final.

50. See A. Rörsch, *Genetic Manipulation in Applied Biology*, study contract 346–77–7 ECI NL, EUR 6078, 1976, Commission of the European Communities and D. Thomas, *Production of Biological Catalysts, Stabilization and Exploitation*, study contract 345–77–6 ECI F, EUR 6079, 1979, Commission of the European Communities.

51. See R. Walgate, 'Brussels Asked to Spend £16 million on Biotechnology', *Nature*, Vol. 283, 10 January 1980, p.125. This is a part of a larger report in that issue on biotechnology, which owes much to the studies by Rörsch and Thomas. Japan did, however, lack genetic engineers.

52. See 'European Biotechnology Lies in Disarray', *Nature*, Vol. 290, 16 April 1981, pp.535–6 and S. Yanchinski, 'Biotechnology in the Swim', *New Scientist*, 30 April 1980.

53. See K. Sargeant, 'Biotechnology in Europe', in *Biotech 83, op. cit.*, p.33, Box 3.

54. For a brief review of the new programme, see 'EEC's Blitz on Biotechnology', in *Biobulletin*, Vol. 2, No. 1, the newsletter of the Science and Engineering Research Council, Swindon, June 1985. See also A. Lubinska, 'European Biotechnology', *Nature*, Vol. 314, 28 March 1985, p.309.

55. See C. Schuuring, 'Dutch Recombinant DNA Guidelines to be Relaxed', *Nature*, Vol. 273, 29 June 1978, p.698, and J. Becker, 'DNA Research Guidelines: Dutch Get Tough', *Nature*, Vol. 290, 9 April 1981, p.436. See G. Maslen, 'Genetic Engineering Debate Reaches Australia', *New Scientist*, 5 April 1979, p.6, and B. Lee, 'Genetic Engineering Down Under', *New Scientist*, 24 July 1980, p.270. For information on Eastern bloc controls see Cohen *et al*, *op. cit.* See P. Newark, 'WHO

Looks for Benefits from Genetic Engineering', *Nature*, Vol. 272, 20 April 1978, pp.663–4. For other information on the activities of the WHO regarding genetic manipulation, see documents in the MIT Archives, and Appendices of the *Report of the Federal Interagency Committee, op. cit.* See Bull *et al., op. cit.*

CHAPTER 6

1. J. Bronowski, *The Ascent of Man*, BBC Publications, London, 1973, p.390.
2. J.D. Watson, 'Let Us Stop Regulating DNA Research', *Nature*, Vol. 278, 8 March 1979, p.113.
3. See S. Yanchinski, *Setting Genes to Work: The Industrial Era of Biotechnology*, Penguin, Harmondsworth, 1985, pp.113–14.
4. See, for example, A. Irwin, D. Smith and R. Griffiths, 'Risk Analysis and Public Policy for Major Hazards', *Physics and Technology*, Vol. 13, No. 6, 1982, pp.258–65.
5. See S. Krimsky, *Genetic Alchemy: A Social History of the Recombinant DNA Controversy*, MIT Press, Cambridge, Mass., 1982.
6. *E. coli* K-12 was the main enfeebled strain used as an EK1 organism under NIH guidelines.
7. See RAC, minutes of meeting, 15–16 January 1977, MIT Archives. Enteric biologists and gastroenterologists were to be included.
8. A.M. Skalka, 'Report from COGENE's Observer at the Falmouth Workshop on Risk Assessment', in A.M. Skalka *et al., First Report to COGENE from the Working Group on Risk Assessment*, ICSU, July 1978, pp.16–20. See also the official report by S.L. Gorbach, 'Recombinant DNA: An Infectious Disease Perspective', *Journal of Infectious Diseases*, Vol. 137, 1978, pp.615–23. The formal report took a year to publish, but in the meantime an influential letter was sent by Gorbach to D.S. Fredrickson emphasising the consensus. See National Institutes of Health, *Environmental Impact Statement on NIH Guidelines for Research Involving Recombinant DNA Molecules*, Part Two, US Department of Health, Education and Welfare, Bethesda, Maryland, 1977, Appendix M.
9. Krimsky does, however, acknowledge the more circumspect report produced by Gorbach over a year after the letter.
10. US Department of Health, Education and Welfare, NIH, 'US-EMBO Workshop to Assess Risks for Recombinant DNA Experiments Involving the Genomes of Animal, Plant and Insect Viruses', *Federal Register*, 31 March 1978, Part III. Note than an earlier NIH/EMBO workshop held in March 1977 examined the parameters of physical containment.
11. See European Science Foundation, *Report 1978*, Strasbourg, p.37.
12. See US Department of Health, Education and Welfare, NIH, 'Proposed Revised Guidelines on Recombinant DNA Research', *Federal Register*, 27 September 1977.

13. See US Department of Health, Education and Welfare, NIH, 'Recombinant DNA Research: Proposed Revised Guidelines', *Federal Register*, 28 July 1978. For a summary of the criticisms see N. Wade, 'Gene Splicing Rules: Another Round of Debate', *Science*, Vol. 199, 6 January 1978, p.30. See also Krimsky, *op. cit.*, pp.233–43. The lack of an environmental impact assessment with the 1977 proposed revisions was a notable criticism, over which Friends of the Earth threatened legal action. See R.M. Harzman, letter to D.S. Fredrickson, NIH, 30 September 1977, on behalf of Friends of the Earth, in J.D. Watson and J. Tooze, *The DNA Story: A Documentary History of Gene Cloning*, W.H. Freeman & Co., San Francisco, 1981, pp.266–7. The July 1978 proposed revisions included such a document. For a summary of these revisions, see D. Dickson, 'US Proposes Exemptions from DNA Guidelines', *Nature*, Vol. 274, 10 August 1978, p.411.

14. Rowe had been centrally involved in risk assessment work and had also questioned the wisdom of the initial calls for caution, claiming that the range of expertise of those scientists who did so was too narrow. See Krimsky, *op. cit.*, pp.234–5 and Anon., 'Scientists Debate Safety of Research on *E. coli* Strain', *Nature*, Vol. 279, 31 May 1979, p.360.

15. R. Curtiss, letter to D.S. Fredrickson, NIH, 4 October 1979, reprinted in Watson and Tooze, *op. cit.*, pp.446–7. He referred to Falmouth data. Additional data came from a 'worst-case' experiment carried out by Rowe and a colleague, Malcolm Martin, similar to the experiment Berg had planned years earlier, which had sparked off the whole issue.

16. Many of the letters for and against the exemptions, or critical of the voting procedure, are reproduced in Watson and Tooze, *op. cit.*, pp.438–59.

17. See A. Campbell, 'Natural Modes of Genetic Exchange and Change', in J. Morgan and W.J. Whelan (eds), *Recombinant DNA and Genetic Experimentation*, Pergamon Press, Oxford, New York, 1979, pp.21–7.

18. 'Genetic Manipulation: New Guidelines for UK', *Nature*, Vol. 276, 9 November 1978, pp.104–8. For a more complete description of the reasoning behind the new approach, see GMAG, *Second Report of the Genetic Manipulation Advisory Group*, HMSO, London, Cmnd 7785, 1979. The HSE also undertook its own consultations on the proposals.

19. GMAG, *Second Report, op. cit.*, p.31. See also E. Lawrence, 'Recombinant DNA Hazards May Be Reassessed', *Nature*, Vol. 274, 20 July 1978, p.203, and Anon. 'Rational Risk Assessment for Genetic Engineering', *New Scientist*, 9 November 1978, p.421.

20. 'Now Reason Can Prevail', editorial, *Nature*, 9 November 1978, p.103.

21. For an attempt to assign probabilities to a supposedly hazardous experiment, see R. Holliday, 'Should Genetic Engineers Be Contained?', *New Scientist*, 17 February 1977, pp.399–401. For a cancer-causing epidemic arising from a shotgun experiment (perhaps involving human DNA as donor) Holliday estimated a total risk of 10^{-14}. He estimated a risk of 10^{-11} for a single death or, as he puts it: if ten scientists in each of 100 laboratories carried out 100 experiments per year, the likelihood of the least tragic result (a single death) would occur

on average once a million years. See also D.R. Lincoln, L.R. Landis and H.A. Gray, 'Containing Recombinant DNA: How to Reduce the Risk of Escape', *Nature*, Vol. 281, 11 October 1979, pp.421-2.

22. For a comprehensive discussion of these sorts of issues in relation to a number of examples of actual disasters, see B.A. Turner, *Man-Made Disasters*, Wykeham Publications, London, 1978.

23. See W. Barnaby, 'Swedish Firm in Recombinant DNA Storm', *Nature*, Vol. 282, 15 November 1979, p.222.

24. S. Falkow, letter to D.S. Fredrickson, NIH, 19 April 1978, in Watson and Tooze, *op. cit.* pp.344-5. The proceedings of this conference have been published as H. Boyer and S. Nicosia (eds), *Genetic Engineering*, Elsevier/North Holland Biomedical Press, Amsterdam, New York, Oxford, 1978.

25. Sir John Kendrew in evidence, in the House of Lords, Select Committee on the European Communities, *Genetic Manipulation (DNA)*, HMSO, London, HL 188, 1980, p.22.

26. This comment was made with the qualification that the authors supported GMAG's policy of not relaxing or abandoning controls without a thorough assessment of risk and benefit. See Advisory Council for Applied Research and Development, Advisory Board for the Research Councils and the Royal Society, *Biotechnology: Report of a Joint Working Party*, HMSO, London, 1980 (known as the Spinks Report), p.33.

27. The WHO had suggested that COGENE might develop such international guidelines. See Draft Minutes of the Fourth Meeting of COGENE, May 1980.

28. R. Curtiss, 'Comments on Callahan', in J. Richards (ed.), *Recombinant DNA: Science, Ethics and Politics*, Academic Press, New York, London, 1978, p.154.

29. Particularly good coverage of the US experience comes from Krimsky, *op. cit.* and I.H. Carmen, *Cloning the Constitution: An Inquiry into Governmental Policymaking and Genetic Experimentation*, University of Wisconsin Press, Madison, London, 1985.

30. See B.K. Zimmerman, 'Beyond Recombinant DNA–Two views of the Future', in Richards (ed.), *op. cit.*, pp.272-301.

31. See Watson and Tooze, *op. cit.*, pp.140-1.

32. In both cases if violation was wilful the fine could be replaced or supplemented by a one-year prison sentence.

33. See D. Dickson, 'Friends of DNA Fight Back', *Nature*, Vol. 272, 20 April 1978, pp.664-5.

34. R. Curtiss, letter to D.S. Fredrickson, NIH, (with copies to the various committees considering legislation) 12 April 1977, MIT Archives. Curtiss still retained some caution, however, as shown by his subsequent criticism of US procedures at the 1979 revisions. For a general description and a collection of related documents, see Watson and Tooze, *op. cit.*, pp.137-201. For an analysis of the course of US legislation and its links with risk assessment and the lobbying of scientists, see Krimsky, *op. cit.*, pp.312-38. For a comparison between

US moves to produce legislation and the UK use of existing legislation, see S. Wright, 'Molecular Politics in Great Britain and the United States: The Development of Policy for Recombinant DNA Technology', *Southern California Law Review*, Vol. 51, No. 6, September 1978, pp.1396–400.

35.   N. Zinder, 'The Gene Scientist and the Law', in Watson and Tooze, *op. cit.*, p.199. N.B. other US agencies subsequently examined the possibility of extending regulations within their remits; for example, the Food and Drugs Administration and the Environmental Protection Agency.

36.   See *Science Policy Implications of DNA Recombinant Molecule Research*. Hearings before the Subcommittee on Science, Research and Technology, US House of Representatives, 95th Congress, First Session, US Government Printing Office, Washington DC, 1977. In particular, see the testimony by J.G. Adams, Vice-President, Scientific and Professional Relations, Pharmaceutical Manufacturers' Association, and I.S. Johnson, Vice-President of Research, Lilly Research Laboratories, pp.371–490.

37.   In the 1980s new regulatory measures have been considered with the growth in requests for the RAC to approve the release of genetically-altered organisms into the environment.

38.   See N. Wade, 'Recombinant DNA: NIH Rules Broken in Insulin Gene Project', *Science*, Vol. 197, 30 September 1977, pp.1342–44.

39.   See C. Marwick, 'Genetic Engineers in the Sin-bin...', *New Scientist*, 11 September 1980, p.764.

40.   See D. Dickson, 'NIH Censure for Dr. Martin Cline', *Nature*, Vol. 291, 4 June 1981, p.369. See also, 'Furore Over Human Genetic Engineering', *New Scientist*, 16 October 1980, p.140.

41.   R. Curtiss, letter to D.S. Fredrickson, NIH, 12 April 1977, *op. cit.*, p.13. For a general assessment of the applicability and enforceability of guidelines in different states, see S.N. Cohen *et al.*, *Report to COGENE from the Working Group on Recombinant DNA Guidelines*, Committee on Genetic Experimentation, International Council of Scientific Unions, Miami, 1980, pp.38–42.

42.   See E. von Weizsäcker, 'The Environmental Dimensions of Biotechnology', in D. Davies (ed.), *Industrial Biotechnology in Europe: Issues for Public Policy*, Frances Pinter, London and Dover, NH, 1986, pp.35–45.

CHAPTER 7

1.   See, for example, US Congress, Office of Technology Assessment, *Commercial Biotechnology: An International Analysis*, US Government Printing Office, 1984. On the UK see Advisory Council for Applied Research and Development, Advisory Board for the Research Councils and the Royal Society, *Biotechnology: Report of a Joint Working Party*, HMSO, London, 1980 (also known as the Spinks Report), and P.

Dunnill and M. Rudd, *Biotechnology and British Industry*, a report to the Biotechnology Directorate of the Science and Engineering Research Council, Swindon, 1984. See also 'France Unveils a Grand Plan for Biotechnology', *New Scientist*, 22 July 1982, p.212. For a summary of Japanese policy, see M. Tanaka (Director of the Bioindustry Office at the Japanese Ministry of International Trade and Industry), 'Biotechnology in Japan', in *Biotech 83: Proceedings of the International Conference on the Commercial Applications and Implications of Biotechnology*, Online Publications, Northwood, UK, 1983, pp.1–12. See also D. Dickson, 'Biotechnology: Canada Takes Stock', *Nature*, Vol. 290, 9 April, 1981, pp.436–7.

2. See C. Vaughan, 'Systems for Collaboration in Advanced Biotechnology', *Biotech 84: Europe*, Online Publications, Pinner, Middx., 1984, pp.159–73.

3. See R. Walgate, 'Celltech's Deal with MRC', *Nature*, Vol. 289, 26 February 1981, p.737.

4. R. Dietz, 'Fostering Biotechnology in the UK', in *Biotech 84: Europe, op. cit.*, p.35.

5. See G. Potter, *Biobulletin*, Vol. 1, No. 1, SERC, Swindon, May 1984. *Biobulletin* is the official newsletter of the SERC's Biotechnology Directorate.

6. Dunnill and Rudd, *op. cit.*, p.1.

7. See 'France Unveils a Grand Plan for Biotechnology', *op. cit.*, and 'Biotechnology: France Now', *Nature*, Vol. 280, 10 July 1980, p.99. See also P. Douzou, 'A Survey of the French Biotechnology Programme', in *Biotech 83, op. cit.*, pp.21–3.

8. See J.S. Dunnett, 'Biotechnology in West Germany: Brain Drain Threatens Progress', *Nature*, Vol. 302, 21 April 1983, p.644. See also P. Daly, *The Biotechnology Business: A Strategic Analysis*, Frances Pinter, London, 1985, pp.32–3.

9. See G. Gregory, 'Big is Beautiful in Biotechnology', *New Scientist*, 6 December 1984, p.14.

10. See M. Tanaka, 'Biotechnology in Japan', in *Biotech 83, op. cit.*, pp.1–12. Tanaka is the director of the Bioindustry Office at the Ministry. Other important Japanese bodies which are promoting biotechnology include the Agency of Science and Technology, the Ministry of Agriculture, Forestry and Fisheries, the Ministry of Health and Welfare and the Bioindustry Development Centre.

11. See T. Itoh, 'The Features of Biotechnology in Japan', in *Biotech 84: Europe, op. cit.*, pp.19–32.

12. See Daly, *op. cit.*, pp.28–31.

13. See the database for the International Patent Documentation Centre (INPADOC), Vienna, and I. Anderson, 'Gene Giants Line Up for Patent War', *New Scientist*, 1 March 1984.

14. See WIPO, 'Budapest Treaty on the International Recognition of the Deposit of Microorganisms for the Purpose of Patent Procedure', appended to *Report of the Federal Interagency Committee on Recombinant DNA Research: International Activities*, US Department of

Health, Education and Welfare, November 1977. See also Intellectual Property Treaty Series No. 5 (1981), *Budapest Treaty on the International Recognition of the Deposit of Microorganisms for the Purposes of Patent Procedure (with Regulations) Budapest, April 28–December 31, 1977*, HMSO, London, Cmnd 8136.

15. See R. Lawrence, 'Procedures and Pitfalls in Patent Protection', *Biotech 83, op. cit.* pp.121–31, and R. Perry 'What is a Patentable Biological Invention?', *Biotech 84: Europe, op. cit.*, pp.45–56.

16. In the study of international relations in both political and economic terms it is becoming more common to define a semi-developed world in addition to the developed and developing. To significant writers like Immanuel Wallerstein, understanding the international structure of capitalist relations necessitates the inclusion of the 'semi-periphery' as a group of countries with a foot in the developed world, but still dominated by it, which in turn partly dominates the less developed countries.

17. See A.T. Bull *et al., Biotechnology: International Trends and Perspectives*, OECD, Paris, 1982, pp.45–6.

18. E.J. DaSilva, 'Biotechnology – A Helping Hand in Developing Countries', *Biotech 83, op. cit.*, p.43.

19. See M. Kenney, 'Genetic Engineering and Agriculture: Socioeconomic Aspects of Biotechnology R&D in Developed and Developing Countries', *Biotech 83., op. cit.*, pp.475–89.

20. See 'Proposal for Genetic Engineering Centre Upsets India', *New Scientist*, 25 August 1983, p.528, and S. Saraf, 'Let a Hundred Labs Bloom', *Nature*, Vol. 307, 16 February 1984, p.583.

21. See P. Feillet, 'Biotechnology: The Link Between the Food, Chemical and Energy Industries', *Biotech 84: Europe, op. cit.*, pp589–91.

22. On non-tariff barriers in this context, see US Congress, Office of Technology Assessment, *Commercial Biotechnology; An International Analysis*, US Government Printing Office, Washington DC, 1984, p.469. On biotechnology hazards, see K. Sargeant and C.G.T. Evans, *Hazards Involving the Industrial Use of Micro-organisms*, study contract 430–78–5 ECI EUR 6349, 1979, Commission of the European Communities.

CHAPTER 8

1. Where the ESF coordinated policy and the EMBO provided technical support.

2. See D. Crane, 'Transnational Networks in Basic Science', in R.O. Keohane and J.S. Nye (eds), *Transnational Relations and World Politics*, Harvard University Press, Cambridge, Mass., 1972. The UK trade union ASTMS did, however, develop a number of international contacts, for example where trade unions were represented on the Dutch national advisory committee.

3. As opposed to intergovernmental linkages.
4. See F.E. Kast and J.E. Rosenzweig, *Organisation and Management*, McGraw-Hill, Kogusha, London and Tokyo, 1974, p.370.
5. See the evidence of Shirley Williams before the Select Committee on Science and Technology, in *Second Report from the Select Committee on Science and Technology*, Session 1978–79, HMSO, London, 1979, pp.156–70.
6. See S. Murphy *et al., No Fire, No Thunder*, Pluto Press, London and Sydney, 1984. See also M. Levinson, 'Custom-made Biological Weapons', *International Defense Review*, Vol. 11, 1986, pp.1611–15.
7. See C. Joyce, 'New Moves to Control the Splice of Life', *New Scientist*, 25 November 1982, p.486. Conferences which have addressed similar issues have been considered in Chapter 2. The New York Academy of Sciences held one of the few conferences devoted to broader issues in 1975. See M. Lappé and R.S. Morrison (eds), 'Ethical and Scientific Issues Posed by Human Uses of Molecular Genetics', *Annals of the New York Academy of Sciences*, Vol. 265, 23 January 1976.
8. Power is seen here simply as possessing the ability to influence.
9. See J.D. Watson and J. Tooze, *The DNA Story: A Documentary History of Gene Cloning*, W.H. Freeman & Co., San Francisco, 1981, pp.251–61. A number of drafts were considered.
10. *Report of the Working Party on the Practice of Genetic Manipulation*, HMSO, London, Cmnd 6600, 1976, p.13.
11. See S. Wright, 'Molecular Politics in Great Britain and the United States: The Development of Policy for Recombinant DNA Technology', *Southern California Law Review*, Vol. 51, No. 6, September 1978.
12. P. Pringle and J. Spigelman, *The Nuclear Barons*, Sphere Books, London, Holt, Rinehart and Winston, New York, 1981, p.392.

Index